VOLUME ONE HUNDRED AND TWENTY SIX

ADVANCES IN
COMPUTERS

VOLUME ONE HUNDRED AND TWENTY SIX

ADVANCES IN
COMPUTERS

Edited by

ALI R. HURSON
*Missouri University of Science and Technology,
Rolla, MO, United States*

ACADEMIC PRESS

An imprint of Elsevier

Academic Press is an imprint of Elsevier
50 Hampshire Street, 5th Floor, Cambridge, MA 02139, United States
525 B Street, Suite 1650, San Diego, CA 92101, United States
The Boulevard, Langford Lane, Kidlington, Oxford OX5 1GB, United Kingdom
125 London Wall, London, EC2Y 5AS, United Kingdom

First edition 2022

ISBN: 978-0-323-98855-1
ISSN: 0065-2458

For information on all Academic Press publications
visit our website at https://www.elsevier.com/books-and-journals

Publisher: Zoe Kruze
Developmental Editor:
    Cindy Angelita Pe Benito-Gardose
Production Project Manager: James Selvam
Cover Designer: Greg Harris

Typeset by STRAIVE, India

Working together
to grow libraries in
developing countries

www.elsevier.com • www.bookaid.org

Contents

Contributors

Dragan Bojić
School of Electrical Engineering, University of Belgrade, Belgrade, Serbia

Jacques Carette
McMaster University, Hamilton, ON, Canada

Ali R. Hurson
Department of Electrical and Computer Engineering, Missouri University of Science and Technology, Rolla, MO, United States

Roshan P. James
Google, New York, NY, United States

Nenad Korolija
School of Electrical Engineering, University of Belgrade, Belgrade, Serbia

Miloš Kotlar
School of Electrical Engineering, University of Belgrade, Belgrade, Serbia

Jelena Marković
Faculty of Mathematics, University of Belgrade, Belgrade, Serbia

Veljko Milutinović
Department of Computer Science, Indiana University in Bloomington, IN, United States

Marija Punt
School of Electrical Engineering, University of Belgrade, Belgrade, Serbia

Amr Sabry
Indiana University in Bloomington, IN, United States

Jakob Salom
Mathematical Institute of the Serbian Academy of Sciences and Arts, Belgrade, Serbia

Saša Stojanović
School of Electrical Engineering, University of Belgrade, Belgrade, Serbia

Živojin Šuštran
School of Electrical Engineering, University of Belgrade, Belgrade, Serbia

Zhilbert Tafa
University of Montenegro, Podgorica, Montenegro

Aleksandar Veljković
Faculty of Mathematics, University of Belgrade, Belgrade, Serbia

Preface

Traditionally, *Advances in Computers*, the oldest series to chronicle the rapid evolution of computing, annually publishes several volumes, each one typically comprising four to eight chapters, describing new developments in the theory and applications of computing.

The 126th volume is an eclectic one inspired by the recent issues of interest in research and development in computer science and computer engineering. The volume is a collection of five chapters, as follows:

In Chapter 1 entitled "VLSI for SuperComputing: Creativity in R + D from applications and algorithms to masks and chips," Milutinovic et al. report and articulate on their experiences in developing a course that teaches VLSI design in an unconventional manner, to motivate creativity and research within the scope of new computing paradigms, such as those presented in this volume of *Advances in Computers*. The course tries to stress the following issues:

(a) Experiences of the world's top foundries for VLSI in SuperComputing like Qualcomm, Intel, Mubadala, and Toshiba;
(b) Experiences that will teach students worldwide to become competitive for work openings in a new industry; and
(c) Experiences that synergize the different phases of a holistic VLSI design, i.e., from applications to chip fabrication.

In Chapter 2 entitled "Embracing the laws of physics: Three reversible models of computation," Carette et al. question the traditional view of computational models and conventional computer architectures that are based on a macrolevel view of physics and the realization of primitive operations using macrolevel devices. They argue that "this point of view is incompatible with computation realized using quantum devices or analyzed using elementary thermodynamics as both these fundamental physical theories imply that information is a conserved quantity of physical processes and of primitive computational operations." Consequently, they propose to redevelop foundational computational models in a manner that embraces the principle of conservation of information. The chapter defines the information and the notion of conservation in a computational setting, followed by three notions of data and their associated reversible computational models.

Chapter 3 entitled "WSNs in environmental monitoring: Data acquisition and dissemination aspects" by Tafa articulates the need for systematic and continuous identification and management of pollution sources. It further argues that the conventional environmental monitoring systems are based on manual periodic onsite sampling and traditionally using expensive, sparsely deployed instruments. In contrast, one can take advantage of the advances in low-cost sensors, embedded systems and low-power wireless technologies, and the Internet of Things (IoT) information systems as a more feasible approach to environmental monitoring, especially for hardly accessible and harsh environments. The chapter presents, analyzes, and classifies the current technological efforts toward (near) real-time low-cost continuous water and air quality monitoring, focusing mainly on sensing, processing and communication techniques, and algorithms for data acquisition and dissemination.

"Energy efficient implementation of tensor operations using dataflow paradigm for machine learning" by Koltar et al. is the subject of discussion in Chapter 4. The chapter introduces 10 different tensor operations commonly used for handling big data in machine learning and data mining applications, such as deep learning, clustering, classification, dimension reduction, and anomaly detection. The discussion concentrates on the energy-efficient aspect of dataflow implementation of tensor operations. Dataflow implementations of all 10 tensor operations are analyzed comparatively with the related control-flow implementations, based on metrics such as speedup, complexity, power savings, and meantime between failures.

Finally, in Chapter 5 entitled "A runtime job scheduling algorithm for cluster architectures with dataflow accelerators," Korolija et al. focus on task scheduling within the scope of heterogeneous computer cluster architectures for high-performance computing composed of both control-flow and dataflow components. Several scheduling algorithms are proposed and compared against several scheduling algorithms that have advanced in the literature. The chapter attempts to show that the performance of the existing cluster structures enhanced with appropriate dataflow accelerators can considerably improve the overall system performance with a proper scheduling algorithm, while at the same time can reduce the hardware complexity and power consumption.

I hope that the readers will find this volume interesting and useful for teaching, research, and other professional activities. I welcome feedback on this volume as well as suggestions for topics of future volumes.

ALI R. HURSON

Missouri University of Science and Technology, Rolla, MO, USA

CHAPTER ONE

VLSI for SuperComputing: Creativity in R+D from applications and algorithms to masks and chips

Veljko Milutinović[a], Miloš Kotlar[b], Jakob Salom[c], Saša Stojanović[b], Živojin Šuštran[b], Aleksandar Veljković[d], Jelena Marković[d], and Ali R. Hurson[e]

[a]Department of Computer Science, Indiana University in Bloomington, IN, United States
[b]School of Electrical Engineering, University of Belgrade, Belgrade, Serbia
[c]Mathematical Institute of the Serbian Academy of Sciences and Arts, Belgrade, Serbia
[d]Faculty of Mathematics, University of Belgrade, Belgrade, Serbia
[e]Department of Electrical and Computer Engineering, Missouri University of Science and Technology, Rolla, MO, United States

Contents

Abstract

This article describes a course that teaches VLSI design in an unconventional way, to steer creativity and to motivate students to research new computing paradigms, like those presented in this Volume of Advances in Computers.

1. Introduction

This article describes a course that teaches VLSI design in an unconventional way, to steer creativity and to motivate students to research new computing paradigms, like those presented in this Volume of Advances in Computers. The course tries to stress the following issues:

(a) Experiences of the world's top foundries for VLSI in SuperComputing like Qualcom [1], Intel [2], and Mubadala [3]. Some Toshiba experiences were also taken into consideration [4].

(b) Experiences that will teach students worldwide to become competitive for work openings in a new industry.

(c) Experiences that synergize the phases of a holistic VLSI design, namely:
- Phase #1: From Applications to Algorithms,
- Phase #2: From Algorithms to Masks, and
- Phase #3: From Masks to Chips.

In this consideration, Phase #1 is best treated by mathematicians, Phase #2 is best treated by computer engineers, and Phase #3 is best treated by physical chemists. Consequently, versions of this course (dialects) where experimentally offered at three different schools of the Belgrade University: The School of Mathematics, The School of Electrical Engineering, and The School of Physical Chemistry.

In each one of its three dialects, this course is stressing the following three issues:
- Deep professional knowledge,
- Detailed multi-dimensional verification, and
- Relevant inter-disciplinary management.

The rest of this article describes the major elements of the presented course. These elements are the same for each one of its dialects; what differs are the teaching examples and homework assignments.

The rest of this article concentrates on the course dialect of interest to electrical engineering; so, the stress is on the transformation from algorithms to masks. In other words, the main question is, for a given algorithm, what is the best implementation of the corresponding VLSI chip. For that question to be answered effectively, an effort has to be invested into the induction of creativity among students, so a part of this text concentrates on that issue, too.

This course is different than any other VLSI design course that the authors are aware of, because it compares different paradigms (all major paradigms nowadays) and enables students to get the selected four algorithms (from four different application areas) through all the VLSI computing paradigms covered.

The four topics covered are: mathematics, image understanding, machine learning, and tensorflow. The four different paradigms covered are: (a) controlflow (multicore vs manycore), (b) dataflow (Maxeler DFE vs Google TPU), (c) diffusionflow (IoT vs WSN), and (d) advanced paradigms (quantum vs biological computing).

For the first two paradigms with four sub-paradigms in total, students are shown examples from mathematics, image understanding, machine learning, and then their task is to play with changes of relevant parameters, to study the effects.

Another important aspect of this course is that it also teaches the techno-economical aspects. It covers: (a) the methodologies to create ideas about possible improvements of the existing state-of-the-art (Blagojevic K20, Bankovic M24), (b) the mechanisms for fund-raising needed to implement the ideas, (c) the approaches of interest for project planning (CMMI and Scrum), and (d) the essence of patenting and trademarking.

Students appreciate techno-economical issue equally as technological issues. One of the past students currently holds the intellectual property of the largest patent settlement in the history of the planet for ICT (Kavcic ref); another one was involved in one of the largest Oracle acquisitions ever (Endeca), still another one is a vice-president of Qualcomm at the time this paper was written (Milivoj Aleksic), while yet another one was a vice-president of Intel (Dado Vrsalovic).

2. From algorithms to implementations

The electrical engineering dialect of the course is divided into three parts:
- Part #1: VLSI for ControlFlow SuperComputing,
- Part #2: VLSI for DataFlow SuperComputing, and
- Part #3: VLSI for WirelessFlow SuperComputing.

Part #1 treats two topics of importance for ManyCore Systems and two topics of importance for MultiCore Systems, as follows:

```
ManyCore Systems:
    VHDL vs Verilog (0.5 weeks)
        Design and programming of a 200 MHz microprocessor (2.5 weeks)
MultiCore Systems:
    MicroProcessor and MultiMicroProcessor systems (1 week)
        Testing    and verification (2 weeks)
```

Part #1 is based on Refs. [5,6]; also, on standard references related to VHDL and Verilog, plus verification and testing. The first homework of the course is to use VHDL to describe a processor that is similar to one core of the

NVidia Tesla system. The first lab of the course is based on the testing and verification approach of ELSYS [7].

Part #2 treats two topics of importance for FineGrain DataFlow Systems and two topics of importance for SystolicArray DataFlow Systems, as follows:

```
FineGrain DataFlow Systems:
                Altera vs. Xilinx (0.5 weeks)
                Design and programming of a 200 MHz Maxeler Machine
(3.5 weeks)
SystolicArray DataFlow Systems:
        SystolicArray architectures (0.5 week)
        A systolic architecture by DARPA (0.5 weeks)
```

Part #2 is based on selected parts of Refs. [8,9]; also, on standard references related to Altera, Xilinx, DataFlow (Maxeler), and a systolic architecture developed by DARPA [10]. The second homework of the course is to program a Maxeler application. The second lab of the course is to study the details of the DARPA's implementation of the Gram-Schmidt orthogonalization algorithm [9] and to reimplement it is using the Google TPU machine.

Part #3 treats two topics of importance for WirelessFlow SuperComputing system design and two topics of importance for the application thereof, as follows:

```
WirelesFlow SuperComputing design:
                Hardware (0.25 weeks)
                Software (0.25 weeks)
WirelessFlow SuperComputing applications:
        Ubiquitous computing with Wireless Sensor Networks and IoT
(0.25 weeks)
                DataMining from Wireless Sensor Networks and IoT
        (0.25 weeks)
```

Part #3 is based on selected topics from Refs. [11,12]. Homework assignment and lab for this part are optional (for a higher grade).

3. Teaching experiences

The bottom line of this course is bringing advanced industrial experience into the classroom. In the first part, the experience is oriented to DARPA's first 200 MHz GaAs microprocessor. In the second part, the experience is oriented to the currently most successful Data Flow supercomputer—Maxeler. In the third part, the experience is oriented to the EU FP7 project ProSence.

Behavior of the students is observed for three different phases of their professional lifetime:

(a) During the course,

(b) During the first decade after the graduation, and

(c) Around the peak of their professional lifetime, which is beyond the first decade of their professional life.

During the semester, students complain that the course is difficult. It is rather hard for them to accept the complex content of the course—quite a lot of topics with their in–depth analysis. Also, a lot of time is needed to successfully complete homework assignments and lab examples. For not just a few of them this course was the most demanding 1 during all 4 years of bachelor studies.

During the first decade after the graduation, they typically say that the knowledge from this course helped them impress interviewers at job application processes. Also, the good knowledge of computer design gave them an edge in their overall IT working engagement.

Some of them had chosen VLSI to be their major life orientation. At the peak of their professional life, they say that this course created for them a decisive advantage when they were trying to acquire the highest professional positions in the major VLSI industries. One of the students of this course later became a VP of Intel. One of the TAs of this course later became a VP of Qualcomm.

4. Creativity in computing

After the teaching, lab exercises, and homework assignments, students are invited to join research efforts, after appropriate preparations.

The preparations include two phases:

(a) Teaching the methods to enhance creativity [13,14], and

(b) Teaching the methods to present research results [15,16].

Then, the students are led to create the following conclusions:

(a) If the communication delays are negligible, the control flow approach could be the optimal paradigm.

(b) If the communication delays are massive, but data movement distances could be made short, the dataflow approach could be the optimal paradigm.

(c) If the communication delays are massive, but data movement distances cannot be made shorter, then the diffusion flow approach could be the optimal one.

Consequently, since the communication delays, relatively to ALU delays, are getting higher and higher, the stress in the related class-research is on the ultimate potentials on the data-flow approach, to be referred next as the "ultimate data-flow."

5. The ultimate DataFlow

The recent public talks and university courses of Veljko Milutinovic were concentrating on the concept of Ultimate DataFlow for BigData, its potentials (up to 2000 in speed up, up to 200 in transistor count, and up to 20 in power savings), and its essence. Consequently, this section covers the issues related to the potentials of the concept, using the programming method of Maxeler, which is still far away from the ideal Ultimate DataFlow, but does achieve considerable speedups over Intel, and thus is of interest for the educational mission described here.

For power, the ratio $20\times$ was quoted, because Intel works on about 4 GHz and current FPGAs on about 200 MHz, which makes about $20\times$.

For transistor count, the ratio of $200\times$ was quoted, for the following reason: If one looks up the Intel microprocessor floorplan one finds out that only 0.5% of the area is for Arithmetic and Logic, making the $1/200\times$ ratio.

For speedup, the frequently quoted numbers are: (a) $20\times$ as the lowest number on Maxeler speedup in recent publications at prestigious journals, (b) $200\times$ as the highest number ever reported by Maxeler, at a respectable publication, and (c) $2000\times$ for the reason, that has nothing to do with existing DataFlow implementations, but has a lot to do with Ultimate DataFlow. (d) even $20,000\times$ could be hoped for some apps!

In Ultimate DataFlow, the speedup depends predominantly on the contribution of loops to the overall execution time:

- If loops contribute with more than 99.95% to the overall run time, then one can hope for a speedup of $2000\times$.
- If one looks up some of the applications on the list of current dataflow successes, one finds out that in many cases the contribution of loops was well over 99.995%, which is why the potentials of Ultimate DataFlow could reach even $20,000\times$.

The fact is that about 4000 students world-wide have used the Maxeler MISANU dataflow machine (https://maxeler.mi.sanu.ac.rs/), and that these students come from universities like: MIT, Harvard, Princeton, Yale, Columbia, NYU, Purdue, University of Indiana in Bloomington, etc.

(in USA), ETH, EPFL, UNIWIE, TUWIEN, Karlsruhe, Heidelberg, Manchester, Bristol, etc. (in Europe), and of course from leading schools of Belgrade: ETF, MATF, FON, FFH.

The Ultimate DataFlow, as a concept, is built on the following two premises (each one with four sub-premises):

1. *Compiler does the following:*

   (a) Separates effectively spatial and temporal data, to satisfy the requirements of the Nobel Laureate Ilya Prigogine, since that action lowers the entropy of a computer system, meaning that the rest of the compiler could do a much better optimization job (lower entropy brings more order and consequently better optimization opportunities).

   (b) Maps the execution graph in the way that makes sure that edges are of the minimal length, to be consistent with the observations of Nobel Laureate Richard Feynman.

   (c) Enables one to go to a lower precision, for what is not of ultimate importance, and consequently to save on resources, that could be reinvested into what *is* of ultimate importance, following the wisdom of Nobel Laureate Daniel Kahneman.

   (d) Enables one to trade between latency and precision, which, in latency-tolerant applications, brings more precision with less resources, and in latency-intolerant applications, brings less latency, in exchange for a lower precision, thus following the wisdom of Nobel Laureate Andre Geim to be implemented.

Unfortunately, none of the dataflow compilers, as far as we know, does any of the above.

2. *Hardware consists of the following:*

   (a) Analog data path of the honeycomb structure, to which one could effectively map the execution graphs corresponding to loops [17].

   (b) The DataPath clocked at a much lower frequency, and hopefully not clocked at all, if the analog path is not unacceptably long, so it is literally the voltage difference between input and output, that moves data thru the graph, as in Ref. [18].

   (c) Digital memory is on the side of the DataPath, so that computing parameters could be kept non-volatile, and temporary results could be stored more effectively.

   (d) The I/O connecting the host and the dataflow is much faster.

Unfortunately, FPGAs offer none of the above today! Consequently, FPGA is today only the least bad solution on the road to the ultimate goal!

6. Conclusion

The major purpose of this course was to lead students through an independent design of the computing infrastructure in the core of three different computing paradigms: ControlFlow, DataFlow, and WirelessFlow. By doing designs, through the standard mechanism of homework assignments, the students were able not only to learn the state of the art, but also to create backgrounds of importance for their inventive reach out into the world of computer engineering research.

This approach is of benefit to those students that plan to join the research groups of the major industrial plants, where creativity needed to generate new concepts is of crucial importance. That is why a certain number of teaching hours, semester by semester, was given to industrial experts who would share their lifetime professional experiences. In later offerings of this course, it was the former students who would come back to teach as industrial experts. Two classes were taught by two different Nobel Laurates, Martin Perl on Creativity [19] and Jerome Friedman on Inventively [20]. Their speeches were a part of two large conferences, one to celebrate 200 years of the Belgrade University (2008), and the other one to enhance the World Rector Conference held in Belgrade (2014).

The course opened a number of new problems for future research, namely: (A) How to maximize the level of creativity of the students that attend the course, both during the process of the work on homework assignments and during the process of the work in industries worldwide; (B) How to maximize the level of understanding of essential issues behind the paradigm, so students could generate new code which is as effective (scalable) and as efficient (fast) as absolutely possible; and (C) How to maximize the quality of their skills to pass the absorbed knowledge to the next generation of students.

Acknowledgment

The authors are thankful to professors Michael Flynn of Stanford and Oskar Mencer of Imperial for extremely valuable comments related to the improvement of the course describes in this article. Also, to Nobel Laurates Martin Perl of Stanford and Jerome Friedman of MIT, for their valuable discussing during their visits to Belgrade.

References

[1] L. Meng, D. Nagalingam, C.S. Bhatia, A.G. Street, J.C.H. Phang, Distinguishing morphological and electrical defects in polycrystalline silicon solar cells using scanning

electron acoustic microscopy and electron beam induced current, Sol. Energy Mater. Sol. Cells 95 (9) (2011) 2632–2637.

[2] D.E.E. Rodriguez, A.G. Ávila Bernal, Development of a bottom–up compact model for Intel®'s High-K 45 nm MOSFET, in: IAENG Transactions on Engineering Technologies, Springer, Netherlands, 2013, pp. 123–134.

[3] A.S. Al-Hammadi, M.E. Al-Mualla, R.C. Jones, Transforming an Economy Through Research and Innovation, vol. 185, University Research for Innovation, 2010.

[4] S. Watanabe, K. Tsuchida, D. Takashima, Y. Oowaki, A. Nitayama, K. Hieda, H. Takato, et al., A novel circuit technology with surrounding gate transistors (SGT's) for ultra high density DRAM's, IEEE J. Solid State Circuits 30 (9) (1995) 960–971.

[5] V. Milutinovic, Surviving the Design of a 200'MHz RISC Microprocessor: Lessons Learned, IEEE Computer Society Press, Los Alamitos, California, USA, 1997.

[6] V. Milutinovic, Microprocessor and Multimicroprocessor Systems, Copyright by Wiley, 2000.

[7] R. Vemuri, R. Kalyanaraman, Generation of design verification tests from behavioral VHDL programs using path enumeration and constraint programming, IEEE Trans. Very Large Scale Integr. VLSI Syst. 3 (2) (1995) 201–214.

[8] V. Milutinović, J. Salom, N. Trifunovic, R. Giorgi, Guide to DATAFLOW SUPERCOMPUTING, Springer, 2015.

[9] V. Milutinovic (Ed.), High-Level Language Computer Architecture, Freeman Computer Science Press, Rockville, Maryland, 1989. Foreword: M. Flynn (Stanford), Turing laureate, 474 p.

[10] H. Vlahos, V. Milutinovic, GaAs microprocessors and digital systems: an overview of R&D efforts, IEEE Micro 8 (1) (1988) 28–56.

[11] L. Gavrilovska, S. Krco, V. Milutinovic, I. Stojmenovic, R. Trobec, Application and multidisciplinary aspects of wireless sensor networks, in: Concepts, Integration, and Case Studies, Computer Communications and Networks, first ed., 2011, ISBN: 978-1-84996-509-5. X, 282 p., Hardcover.

[12] Z.B. Babovic, J. Protic, V. Milutinovic, Web performance evaluation for internet of things applications, IEEE Access 4 (2016) 6974–6992.

[13] V. Milutinovic, A structured approach to research for PhD students in computer science and engineering: how to create ideas, conduct research, and write papers, IPSI BGD Trans. Internet Res. 11 (2015) 47–54.

[14] V. Blagojević, et al., A systematic approach to generation of new ideas for PhD research in computing, in: Advances in Computers, vol. 104, Elsevier, 2017, pp. 1–31.

[15] V. Milutinovic, The best method for presentation of research results, IEEE TCCA Newslett. (1996) 1–6.

[16] V. Milutinović, A good method to prepare and use transparencies for research presentations, IEEE TCCA Newslett. (1997) (online) Available from: http://tab.computer.org/tcca/news/sept96/bestmeth. pdf [acessed 16 June 2011].

[17] V.E.L.J.K.O. Milutinovic, Mapping of neural networks on the honeycomb architecture, Proc. IEEE 77 (12) (1989) 1875–1878.

[18] V. Milutinovic, D. Fura, W. Helbig, An introduction GaAs microprocessor architecture for VLSI, in: Computer;(United States), 1986. 19.3.

[19] M.L. Perl, Developing creativity and innovation in engineering and science, Int. J. Mod. Phys. A 23 (27n28) (2008) 4401–4413.

[20] J.I. Friedman, P. Galison, S. Haack, B.E. Frye, The Humanities and the Sciences, vol. 47, American Council of Learned Societies, 2000.

About the authors

Prof. Veljko Milutinović (1951) received his PhD from the University of Belgrade in Serbia, spent about a decade on various faculty positions in the USA (mostly at Purdue University and more recently at the University of Indiana in Bloomington), and was a co-designer of the DARPAs pioneering GaAs RISC microprocessor on 200MHz (about a decade before the first commercial effort on that same speed) and was a co-designer also of the related GaAs Systolic Array (with 4096 GaAs microprocessors). Later, for almost three decades, he taught and conducted research at the University of Belgrade in Serbia, for departments of EE, MATH, BA, and PHYS/CHEM. His research is mostly in datamining algorithms and dataflow computing, with the emphasis on mapping of data analytics algorithms onto fast energy efficient architectures. Most of his research was done in cooperation with industry (Intel, Fairchild, Honeywell, Maxeler, HP, IBM, NCR, RCA, etc.). For 10 of his books, forewords were written by 10 different Nobel Laureates with whom he cooperated on his past industry sponsored projects. He published 40 books (mostly in the USA), he has over 100 papers in SCI journals (mostly in IEEE and ACM journals), and he presented invited talks at over 400 destinations worldwide. He has well over 1000 Thomson-Reuters WoS citations, well over 1000 Elsevier SCOPUS citations, and about 4000 Google Scholar citations. His Google Scholar h index is equal to 36. He is a Life Fellow of the IEEE since 2003 and a Member of The Academy of Europe since 2011. He is a member of the Serbian National Academy of Engineering and a Foreign Member of the Montenegro National Academy of Sciences and Arts.

Miloš Kotlar received his B.Sc. (2016) and M.Sc. (2017) degrees in Electrical and Computer Engineering from the University of Belgrade, School of Electrical Engineering, Serbia. He is a PhD candidate at the School of Electrical Engineering, University of Belgrade. His general research interests include implementation of energy efficient tensor implementations using the dataflow paradigm (FPGA and ASIC accelerators) and meta learning approaches for anomaly detection tasks.

Dr Jakob Salom received his BSc degree from the University of Belgrade, School of Electrical Engineering. Author/co-author of three books and dozens of sections in international books. Author/co-author of more than 20 peer-review articles in journals and conferences anthologies.

Saša Stojanović is on the faculty of the Department of Computer Engineering in the School of Electrical Engineering, University of Belgrade, Serbia. His PhD thesis, defended in the year 2015, was related to software similarity. He teaches courses on Embedded Systems and Mobile Devices Programming. His current research is in the fields of Software Similarity and Reverse Engineering.

Živojin Šuštran received the BSc and MSc degrees in electrical engineering and computing from the School of Electrical Engineering, University of Belgrade, Serbia, in 2010 and 2012, respectively, where he is currently pursuing the PhD degree. He is currently a Teaching Assistant with the School of Electrical Engineering, University of Belgrade. He has been involved in the research and development of hardware and software solutions in industry and academia for ten years, with expertise in computer architecture, cache memory design, systems programming, operating systems, and FPGA acceleration. He has coauthored two journal articles and gave talks at conferences in Europe. His current research interests include cache coherence and shared memory algorithms, hardware transactional memory, multicore architectures, and with special emphasis on asymmetric multiprocessors.

Aleksandar Veljković, Teaching Assistant, Faculty of Mathematics, University of Belgrade.

Jelena Marković, Teaching Assistant, Faculty of Mathematics, University of Belgrade.

Ali R. Hurson is a professor of Departments of Computer Science, and Electrical and Computer Engineering at Missouri S&T. For the period of 2008–2012 he served as the computer science department chair. Before joining Missouri S&T, he was a professor of Computer Science and Engineering department at The Pennsylvania State University. His research for the past 35 years has been supported by NSF, DARPA, Department of Education, Department of Transportation, Air Force, Office of Naval Research, NCR Corp., General Electric, IBM, Lockheed Martin, Penn State University, and Missouri S&T. He has published over 330 technical papers in areas including computer architecture/organization, cache memory, parallel and distributed processing, Sensor and Ad Hoc Networks, dataflow architectures, VLSI algorithms, security, Mobile and pervasive computing, database systems, multidatabases, global information sharing processing, application of mobile agent technology, and object-oriented databases.

Professor Hurson has been active in various IEEE/ACM Conferences and has given tutorials on global information sharing, database management systems, supercomputer technology, data/knowledge-based systems, dataflow processing, scheduling and load balancing, parallel computing, pervasive computing, green computing, and sustainability. He served as an

IEEE editor, IEEE distinguished speaker, and an ACM distinguish lecturer. Currently, he is Editor-in-Chief of Advances in Computers, editor of The CSI Journal of Computer Science and Engineering, and editor of Computing Journal.

CHAPTER TWO

Embracing the laws of physics: Three reversible models of computation

Jacques Carette[a], Roshan P. James[b], and Amr Sabry[c]

[a]McMaster University, Hamilton, ON, Canada
[b]Google, New York, NY, United States
[c]Indiana University in Bloomington, IN, United States

Contents

Advances in Computers, Volume 126
ISSN 0065-2458
https://doi.org/10.1016/bs.adcom.2021.11.009

15

Abstract

Our main models of computation (the Turing Machine and the RAM) and most modern computer architectures make fundamental assumptions about which primitive operations are realizable on a physical computing device. The consensus is that these primitive operations include logical operations like conjunction, disjunction and negation, as well as reading and writing to a large collection of memory locations. This perspective conforms to a macro-level view of physics and indeed these operations are realizable using macro-level devices involving thousands of electrons. This point of view is however incompatible with computation realized using quantum devices or analyzed using elementary thermodynamics as both these fundamental physical theories imply that information is a conserved quantity of physical processes and hence of primitive computational operations.

Our aim is to redevelop foundational computational models in a way that embraces the principle of conservation of information. We first define what information is and what its conservation means in a computational setting. We emphasize the idea that computations must be reversible transformations on data. One can think of data as modeled using topological spaces and programs as modeled by reversible deformations of these spaces. We then illustrate this idea using three notions of data and their associated reversible computational models. The first instance only assumes unstructured finite data, i.e., discrete topological spaces. The corresponding notion of reversible computation is that of permutations. We show how this simple model subsumes conventional computations on finite sets. We then consider a modern structured notion of data based on the Curry–Howard correspondence between logic and type theory. We develop the corresponding notion of reversible deformations using a sound and complete programming language for witnessing type isomorphisms and proof terms for commutative semirings. We then "move up a level" to examine spaces that treat programs as data, which is a crucial notion for any universal model of computation. To derive the corresponding notion of reversible programs between programs, i.e., reversible program equivalences, we look at the "higher dimensional" analog to commutative semirings: symmetric rig groupoids. The coherence laws for these groupoids turn out to be exactly the sound and complete reversible program equivalences we seek.

We conclude with some possible generalizations inspired by homotopy type theory and survey several open directions for further research.

1. Reversibility, the missing principle

What kind of operations can computers perform? This question has been answered several times in the last hundred years, where each answer proposes an abstract *model of computation* that specifies allowable operations and (usually) their cost. The emerging consensus, reflected in both early models of computations such as the Turing Machine and the RAM as well as in the early Von Neumann models and in modern computer architectures, is that basic computer operations include logical operations like conjunction, disjunction, and negation, as well as reading from and writing to a large (infinite) collection of memory locations. From this small set of primitive operations emerges all higher level programming languages and abstractions.

No doubt, this consensus on the available primitive physical operations has been successful. Furthermore, these operations *can* indeed be performed on a computer. Yet, today, with a possible quantum computing revolution in sight and an unprecedented explosion in embedded computers and cyber-physical systems, there are reasons to rethink this foundational question. In fact, the calls to rethink this foundational question have been proclaimed by physicists almost forty years ago:

> Mathematical models of computation are abstract constructions, by their nature unfettered by physical laws. However, if these models are to give indications that are relevant to concrete computing, they must somehow capture, albeit in a selective and stylized way, certain general physical restrictions to which all concrete computing processes are subjected.
>
> **Toffoli 1980 [1]**
>
> Another thing that has been suggested early was that natural laws are reversible, but that computer rules are not. But this turned out to be false; the computer rules can be reversible, and it has been a very, very useful thing to notice and to discover that. This is a place where the relationship of physics and computation has turned itself the other way and told us something about the possibilities of computation. So this is an interesting subject because it tells us something about computer rules.
>
> **Feynman 1982 [2]**

These quotes by Toffoli and Feynman both highlight the consequences of two obvious observations: (i) all the operations that a computer performs reduce to basic physical operations; and (ii) there is a mismatch between the logical operations of a typical model of computation (which are logically irreversible) and the fundamental laws of physics (which are reversible). One could certainly dismiss the mismatch as irrelevant to the practice of computing but our thesis is that the next computing revolution is likely to be founded on revised models of computation that are designed to be in closer harmony with the laws of physics.

After a detailed introduction on the origins of *logically reversible computer operations* and an excursion into the origins of *irreversible computer operations*, we will develop in detail three reversible models of computation and discuss their potential applications.

1.1 Maxwell's daemon

To fully appreciate the missing principle of *reversibility* in conventional computing, we go back to an old thought experiment by J.C. Maxwell. The details are codified in a letter that Maxwell wrote to P.G. Tait in 1867— the letter, whose ideas are now known as *Maxwell's daemon*, tells of a thought experiment that seems to indicate that intelligent beings can somehow violate the second law of thermodynamics, thereby violating physics itself. Many resolutions were offered for this conundrum (for a compilation, see the book by Leff and Rex [3]), but none withstood careful scrutiny until the establishment of *Landauer's principle* in 1961 [4]—a principle whose experimental validation happened in 2012 [5].

Maxwell's daemon appears to violate the second law of thermodynamics by having a tiny "intelligence" observing the movement of individual particles of a gas and separating fast moving particles from slow moving ones, thereby reducing the total entropy of the system. Landauer's resolution of the daemon relied on two ideas that had taken root only a few decades earlier: the formal notion of computation (through the work of Turing [6], Church [7], and others) and the formal notion of information (through the work of Shannon [8]). Landauer reasoned that the computation done by the finite brain of the daemon involves getting information about the movement of molecules, storing that information, analyzing that information to act on it, and then—and this is the critical step—overwriting it to make room for the next computation. In other words, the computation that is manipulating information in the daemon's brain *must be thermodynamic work*, thereby bringing the daemon back into the fold of physics.

This is a strange and wonderful idea: information, physics, and computation are inextricably linked. In contrast, when the early models of computation were developed, there was no compelling reason to take the information content of computations into consideration—in fact, at that time there was no quantifiable notion of information. These models followed in the footsteps of logic where, following hundreds of years of tradition, the truth of a statement was seen as *absolute* and independent of any reasoning, understanding, or action. Statements were either true or false with no regard to any *observer* and the idea that statements had information content that should be preserved was outside the classical understanding of logic. Hence the fact that conventional logic operations such as conjunction and disjunction were logically irreversible and hence lose information was not a concern. Landauer's observation implied, however, that ideas in each field have consequences for the other [9–15]. To really appreciate this fact, we delve deeper into the origin of our computational models and argue that they are essentially reflections of contemporary laws of physics.

1.2 Origins of computational models

Current high-level programming languages as well as current hardware are both based on the mathematical formalization of logic developed by

De Morgan, Venn, Boole, and Peirce in the mid to late 1800s. Going back to Boole's 1853 book entitled *An Investigation of the Laws of Thought, on which are Founded the Mathematical Theories of Logic and Probabilities*, we find that the opening sentence of Chapter 1 is:

> The design of the following treatise is to investigate the fundamental laws of those operations of the mind by which reasoning is performed;

which clearly identifies the *source* of the logical laws as mirroring Boole's understanding of human reasoning. A few chapters later, we find:

> **Proposition IV.** That axiom of metaphysicians which is termed the principle of contradiction, and which affirms that it is impossible for any being to possess a quality, and at the same time not to possess it, is a consequence of the fundamental law of thought, whose expression is $x^2 = x$.

This "law" is reasonable in a classical world but is violated by the postulates of quantum mechanics. Although a detailed historical analysis of Boole's ideas in the light of modern physics is beyond our scope, the above quotes should convey the idea that our elementary computing notions date back to ideas that were thought reasonable in the late 1800s.

Machines that "compute" are quite old. Müller (1786) first conceived of the idea of a "difference machine," which Babbage (1819–22) was able to construct. There are other computer precursors as well—the first stored programs were actually for looms, most notably those of Bouchon (1725) which were controlled by a paper tape, and of Jacquard (1804), controlled by chains of punched cards. But it was only in the 20th century that computer science emerged as a formal discipline. One of the pioneering works was Alan Turing's seminal paper [6] of 1936 which established the idea that computation has a formal interpretation and that all computability can be captured within a formal system. Implicit in this achievement however is the idea that abstract models of computation are just that—*abstractions of computation realized in the physical world*. Indeed, going back to Turing's 1936 article *On Computable Numbers, with an Application to the Entscheidungsproblem*, the opening sentence of Section 1 is:

> We have said that the computable numbers are those whose decimals are calculable by finite means [...] the justification lies in the fact that the human memory is necessarily limited.

In Section 9, we find:

> I think it is reasonable to suppose that they can only be squares whose distance from the closest of the immediately previously observed squares does not exceed a certain fixed amount.

It is worth noting that these assumptions are both physical (on distances) and metaphysical (on restrictions of the mind). If we take the human mind to be a physical "machine" which performs computation, then when both of the above assumptions are translated into the language of physics, they embody what is known as the "Bekenstein bound" [16], which is an upper limit on the amount of information that can be contained within a given finite region of space. A detailed historical account of these ideas in the context of modern physics is again beyond our scope. However, the quotes above, like the ones before, should convey the ideas that our theories of computation and complexity are based on some physical assumptions that Turing and others found reasonable in the 1930s.

To summarize, a major achievement of computer science has been the development of abstract models of computation that shield the discipline from rapid changes in the underlying technology. Yet, as effective as these models have been, one must note that they *embody several implicit physical assumptions* and these assumptions are based on a certain understanding of the laws of physics. Our understanding of physics has evolved tremendously since 1900! Thus it is time to revisit these abstractions, especially with respect to quantum mechanics. Indeed one should take the physical principles underlying quantum mechanics, the most successful physical theory known to us, and adapt computation to "learn" from these principles. In the words of Girard [17]:

> *In other terms, what is so good in logic that quantum physics should obey? Can't we imagine that our conceptions about logic are wrong, so wrong that they are unable to cope with the quantum miracle? [...] Instead of teaching logic to nature, it is more reasonable to learn from her. Instead of interpreting quantum into logic, we shall interpret logic into quantum.*

There are, in fact, many different quantum mechanical principles which are at odds with our current models of computation. In this chapter, we will focus on the previously identified principle of *reversibility*. In more detail, we will view data as an explicit representation of *information* and programs as processes that transform information in a reversible way, i.e., processes that are subject to the physical principle of *conservation of information*. We will formalize this idea and follow its consequences, which will turn out to be far reaching.

1.3 Programs as reversible deformations

To better understand the essence of "conservation of information" in the context of computing, we first look for analogous ideas in physics, but this time at the macro scale. Viewing information as a physical object, what does it mean to transform an object in such a way that we do not lose its fundamental character?

For rigid objects (like a chair), the only such transformations are translations and rotations. But what about something more flexible, with multiple representations, such as a water balloon? Such objects can be *deformed* in various ways, but still retain their fundamental character—as long as we do not puncture them or over-stretch them. Ignoring material characteristics (i.e., over-stretching), what is special about these deformations, as well as for translations and rotations, is that they correspond to continuous maps, with a continuous inverse. In fact, even more is true: they are analytic maps, with analytic inverses. For our purpose, the most important part is that such maps are infinitely differentiable. In other words, not only is there an inverse to the deformation, but its derivative is also invertible, and so on.

When we look around, we find many different words for related concepts: isomorphism, equivalence, sameness, equality, interchangeability, comparability, and correspondence, to name a few. Some of these are informal concepts, while others have formal mathematical meaning. More importantly, even among the formal concepts, there are differences—which is why there are so many of them! Because there are many such notions, we also need to walk our way through them to find the one which is "just right." Thus we seek a concept which is neither too strong nor too weak, that will express when some structured information should be treated as "the same." We can draw an analogy with topology: in topology, all point sets can always be equipped with either the discrete or the indiscrete topology, but both of these extremes are rarely useful. We will develop our working notion of "sameness" as we go through the various components that make up a programming language.

Starting from the physical perspective, whatever our notion of data is, we will be interested in programs as representing transformations of that data which are reversible. In other words, we want our programs-as-transformations to "play well" with the inherent notion of "sameness" that our data will carry. Thus we need to start by looking at what structure our data has, which will help us define an appropriate notion of a reversible program. Of course, when programs themselves are data, things do get more complicated. In the following sections, we will look at different natural classes of data, and explore the corresponding notion of reversible programs.

To summarize, we will take "the same" as a fundamental principle and derive what it means for data, programs, program transformations, as well as proofs/deductions, to be "the same"—in a manner consistent with preservation of information. This stands in stark contrast with most current approaches to reversible computation, which start from current models of computation involving irreversible operations and try to find various ways to *patch things up* so as to be reversible.

1.4 Reversible programming languages

The practice of programming languages is replete with ad hoc instances of reversible computations: database transactions, mechanisms for data provenance, checkpoints, stack and exception traces, logs, backups, rollback recoveries, version control systems, reverse engineering, software transactional memories, continuations, backtracking search, and multiple-level undo features in commercial applications. In the early nineties, Baker [13, 18] argued for a systematic, first-class, treatment of reversibility. But intensive research in full-fledged reversible models of computations and reversible programming languages was only sparked by the discovery of deep connections between physics and computation [1, 4, 10, 19, 20], and by the potential for efficient quantum computation [2].

The early developments of reversible programming languages started with a conventional programming language, e.g., an extended λ-calculus, and either

1. extended the language with a history mechanism [21–24], or
2. imposed constraints on the control flow constructs to make them reversible [25].

More modern approaches recognize that reversible programming languages require a fresh approach and should be designed from first principles without the detour via conventional irreversible languages [26–29].

In previous work, Carette, Bowman, James, and Sabry [30–32] introduced the Π family of typed reversible languages. As motivated above, the starting point for this development is the physical principle of *conservation of information* [33, 34] and the family of languages is designed to embrace this principle by requiring all computations to preserve information.

The fragment without recursive types is universal for reversible Boolean circuits [31] and the extension with recursive types and trace operators [35] is a Turing-complete reversible language [30, 31]. While at first sight, Π too might appear ad hoc, it really arises naturally from an "extended" view of the Curry–Howard correspondence [32]: rather than looking at mere *inhabitation* as the main source of analogy between logic and computation, *type equivalence* becomes the source of analogy. Taking inspiration from the fact that many terms of the λ-calculus arise from Cartesian Closed Categories including, most importantly, a variety of propositional equalities and computation rules, allows us to pursue that analogy further. Some of the details of this development will be motivated and explained in the present paper.

2. Data I: Finite sets

Most programming languages provide primitive data like Booleans, characters, strings, and (bounded) numbers that are naturally modeled as finite sets. We therefore start by modeling reversible computations over finite and discrete spaces of points. Infinite sets are more subtle, and will be discussed in the conclusion.

What does it mean to deform a space of points? For example, what transformation can we do on a bag of marbles? Well, we can shuffle them around and that is the only transformation that will preserve the space. Turning to the mathematical abstraction as sets, we ask what does it mean for two finite sets to be "the same"? Well, clearly the sets $A = \{1, 2, 3\}$ and $B = \{c, d\}$ are different. Why? Well, suppose there was a transformation $f : A \to B$ that deformed A into B, and another $g : B \to A$ which undid this transformation. Since f is total, by the pigeonhole principle, two elements of A would be mapped to the same element of B. Suppose that this is 2 and 3, and that they both map to d. But $g(d)$ cannot be both 2 and 3, and so g is not the inverse of f. With just a little more work, we can show that f (and g) must be both injective and surjective. In other words, f (and g) must be a bijection between A and B. And of course this only happens when A and B have the same number of elements. More importantly, given a bijection $f : C \to D$ of finite sets C, D, there always exists another bijection $g : D \to C$ which is f's inverse. So, for finite sets, *bijections* act as reversible deformations.

This discussion is purely "semantic," in the sense that it is about the denotation of simple primitive data (sets) and their reversible deformations (bijections). We would like to reverse engineer a programming language from this denotation. But first, an obvious remark: any two sets C and D of cardinality n are always in bijective correspondence. So we can abstract away from the details of the elements of C and D and instead choose canonical representations—in much the same way as computers choose binary words to represent everything.

Definition 1. For $n \in \mathbb{N}$, denote by $[n]$ the set $\{0, 1, ..., n - 1\}$. We will refer to $[n]$ as the canonical set with n elements.

Bijections on $[n]$ have a specific name: permutations. As is well-known, permutations can be generated by sequential compositions of transpositions. Thus we can create a small language for writing permutations on $[n]$ as:

$$p^n ::= id \mid swap\ i\ j \mid p^n\ ; p^n \tag{1}$$

where $i, j : \mathbb{N}$, $i \neq j$ and $i, j < n$. Note that we could remove id from the language and drop the $i \neq j$ condition so that $swap\ j\ j$ would represent the identity permutation.

For convenience, we write $[2^n]$ for the finite set representing n-bit words with the canonical ordering for binary numbers. Thus when $n = 3$, the finite set has elements $\{0, 1, 2, 3, 4, 5, 6, 7\}$ which correspond to the 3-bit words $\{000, 001, 010, 011, 100, 101, 110, 111\}$. Although this language appears weak, it is universal for reversible Boolean combinational circuits $[2^i] \rightarrow [2^i]$ with i input/output wires.

To illustrate the expressiveness of the language, we develop a few small examples. We start by writing Boolean negation "not" as a permutation $[2^1] \rightarrow [2^1]$, the controlled–not gate (also known as "cnot") as a permutation $[2^2] \rightarrow [2^2]$, and the controlled-controlled-not gate (also known as "toffoli") as a permutation $[2^3] \rightarrow [2^3]$:

$$not = swap\ 0\ 1 \tag{2}$$
$$cnot = swap\ 2\ 3 \tag{3}$$
$$toffoli = swap\ 6\ 7 \tag{4}$$

The "cnot" gate operates on two bits and negates the second (the target bit) if the first one (the control bit) is 1, i.e., it swaps 10 and 11; the "toffoli" gate negates the third bit (the target bit) if both the first two bits (the control bits) are 1, i.e., it swaps 110 and 111.

There is, however, a subtle issue: programming in such an unstructured language is *not* compositional in the sense that using the "not" gate in a larger circuit forces us to change its implementation. Indeed if we had two bits and wanted to use "not" to negate the first bit, we would write the permutation of type $[2^2] \rightarrow [2^2]$ that permutes 00 with 10 *and* permutes 01 with 11, i.e, the permutation $swap\ 0\ 2$; $swap\ 1\ 3$. To illustrate how inconvenient this is, consider the reversible full adder below designed by Desoete et al. [36]:

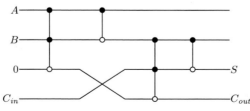

In the figure (copied from a more general paper that includes alternative designs [37]), the full adder takes four inputs: the two bits to add A and B, an incoming carry bit C_{in}, and a heap input initialized to 0 to maintain reversibility. There are four outputs: the first two are identical to the incoming bits A and B and are considered "garbage." The third bit S is the sum and the last bit C_{out} is the outgoing carry bit. In the notation used to describe the circuit, the black dot represent control bits and the white dots represent negations. In our reversible language, we can express this circuit as the following permutation of type $[2^4] \rightarrow [2^4]$:

swap 12 14; *swap* 13 15;	toffoli
swap 8 12; *swap* 9 14; *swap* 10 13; *swap* 11 15;	cnot and swap
swap 6 7; *swap* 14 15;	toffoli
swap 4 6; *swap* 5 7; *swap* 12 14; *swap* 13 15	cnot

$$(5)$$

Note how the implementation of *cnot* as a permutation $[2^2] \rightarrow [2^2]$ cannot be directly reused in the larger circuit $[2^4] \rightarrow [2^4]$.

For such reasons, in programming practice we are interested in structured data and compositional abstractions, which will be the subject of the next section. What we do learn from this short investigation using untyped and unstructured sets is what the *"purely operational"* view of the theory would be. In particular, it tells us that permutations are an inescapable part of the fabric of reversible computing. However as permutations are untyped, and act on the canonicalized version of n-element sets (i.e., those sets where all the structure has been forgotten), these are a rather pale shadow of the rich tapestry of information-preserving transformations of structured data, which we investigate next.

3. Data II: Structured finite types

Instead of spaces (aka discrete sets) consisting solely of unstructured isolated points, we now investigate structured spaces built from sums and products of elementary spaces. This structure corresponds to the building blocks of type theory which are: the empty type ($\perp$), the unit type ($\top$), the sum type ($\uplus$), and the product ($*$) type. Before getting into the formal theory, let us consider possible deformations on the space $(\top \uplus \perp) * (\top \uplus \top)$. This space is the product of two subspaces: the subspace $(\top \uplus \perp)$ which itself is the sum of the space $\top$ containing one element tt and the empty space $\perp$ and the subspace $(\top \uplus \top)$ which is the sum of two spaces each containing the one element tt. First, as discussed in the previous section, any deformation of this space must at least preserve the number of elements: we can neither create nor destroy points during any continuous deformation. Seeing that the number of elements in our example space is 2, a reasonable hypothesis is that we can deform the space above to any other space with 2 elements such as $\top \uplus \top$ or $\top \uplus (\top \uplus \bot)$. What this really means is that we are treating the sum and product structure as malleable. For example, imagining a product structure as arranged in a grid; by "stretching" we can turn it in to a sum structure arranged in a line. We can also change the orientation of the grid by exchanging the axes, as well as do other transformations—as long as we preserve the number of points. Of course, it is not a priori clear that this necessary requirement is also sufficient. Making this intuition precise will be the topic of this section.

3.1 A model of type equivalences

We now want a proper mathematical description of this idea. Our goal is a denotational semantics on types which makes types that have the same number of points be equivalent types. First we note that the structure of types has a nice correspondence (Curry–Howard) to logic:

Logic	Types
false	$\bot$
true	$\top$
$\wedge$	$*$
$\vee$	$\uplus$

This correspondence is rather fruitful. As logical expressions form a commutative semiring, we would expect that types too form a commutative semiring. And indeed they do—at least up to *type isomorphism*. The natural numbers $\mathbb{N}$ are another commutative semiring; it will turn out that, even though the Curry–Howard correspondence has been extremely fruitful for programming language research, it is $\mathbb{N}$ which will be a better model for finite structured types as the corresponding commutative semiring captures the familiar numerical identities that preserve the number of points in the types.

Definition 2. A *commutative semiring* (sometimes called a *commutative rig*—commutative ring without negative elements) $(R, 0, 1, +, \cdot)$ consists of a set R, two distinguished elements of R named 0 and 1, and two binary operations $+$ and $\cdot$, satisfying the following relations for any $a, b, c \in R$:

$$0 + a = a$$
$$a + b = b + a$$
$$a + (b + c) = (a + b) + c$$

$$1 \cdot a = a$$
$$a \cdot b = b \cdot a$$
$$a \cdot (b \cdot c) = (a \cdot b) \cdot c$$

$$0 \cdot a = 0$$
$$(a + b) \cdot c = (a \cdot c) + (b \cdot c)$$

Proposition 1. *The structure* $(\{false, true\}, false, true, \vee, \wedge)$ *is a commutative semiring.*

We would like to adapt the commutative semiring definition to the setting of structured types. First, types do not naturally want to be put together into a "set." This can be fixed if we replace the set R with a universe U, and replace the set membership $0 \in R$ with the typing judgment $\perp : U$ (and similarly for the other items). Our next instinct would be to similarly replace $=$ with a type $A \equiv B$ that asserts that A and B are *propositionally equal*, i.e., reduce to equivalent type-denoting expressions under the rules of the host type system. This is however not true: the proposition $A * B \equiv B * A$ is not normally[a] provable for arbitrary types A and B. But it should be clear that $A * B$ and $B * A$ contain equivalent information. In other words, we would like to be able to witness that $A * B$ can be reversibly deformed into $B * A$, and vice versa, which motivates the introduction of type *equivalences*. To do this, we need a few important auxiliary concepts.

Definition 3 (*Propositional Equivalence*). Two expressions a, b of type A are *propositionally equal* if their normal forms are equivalent under the rules of the host type system.

In Martin-Löf Type Theory, normal forms mean $\beta\eta$-long normal forms under α-equivalence. In other words, expressions are evaluated as much as possible (β-reduced), all functions are fully applied (η-long), and the exact names of bound variables are irrelevant (α-equivalence). Note that the above definition applies equally well to expressions that denote values and expressions that denote types.

---

[a] Except in univalent type theory where equivalent types are identified.

Definition 4 (*Homotopy*). Two functions $f, g : A \to B$ are *homotopic* if $\forall x : A. f(x) \equiv g(x)$. We denote this $f \sim g$.

It is easy to prove that homotopies (for any given function space $A \to B$) are an equivalence relation. The simplest definition of the data which makes up an equivalence is the following.

Definition 5 (*Quasi-inverse*). For a function $f : A \to B$, a *quasi-inverse* is a triple (g, α, β), consisting of a function $g : B \to A$ and two homotopies $\alpha : f \circ g \sim \mathrm{id}_B$ and $\beta : g \circ f \sim \mathrm{id}_A$.

Definition 6 (*Equivalence of types*). Two types A and B are equivalent $A \simeq B$ if there exists a function $f : A \to B$ together with a quasi-inverse for f.

Why *quasi*? The reasons are beyond our scope, but the interested reader can read Section 2.4 and Chapter 4 in the Homotopy Type Theory (HoTT) book [38]. There are several conceptually different, but equivalent, "better" definitions. We record just one here:

Definition 7 (*Bi-invertibility*). For a function $f : A \to B$, a *bi-inverse* is a pair of functions $g, h : B \to A$ and two homotopies $\alpha : f \circ g \sim \mathrm{id}_B$ and $\beta : h \circ f \sim \mathrm{id}_A$.

We can then replace quasi-inverse with bi-invertibility in the definition of type equivalence. The differences will not matter to us here.

We are now in position to describe the commutative semiring structure for types. After replacing the set R with a universe U, we also replace the algebraic use of $=$ in Definition 2 by the type equivalence relation $\simeq$. With this change, we can indeed prove that types (with $\bot$, $\top$, $\uplus$, $*$) form a commutative semiring. The reader familiar with universal algebra should pause and ponder a bit about what we have done. We have lifted *equality* from being in the signature of the ambient logic and instead put it in the signature of the algebraic structure of interest. In simpler terms, we shift equality from having a privileged status in our meta-theory, to being just another symbol (denoting an equivalence relation) in our theory. The understanding that equality is not an absolute concept has recently been an area of active research in mechanized mathematics—although the concepts of intensional versus extensional equality go back to Frege and Russell.

If we revisit the Curry–Howard correspondence, we notice one more issue. In logic, it is true that $A \vee A = A$ and $A \wedge A = A$. However, neither $A \uplus A$ nor $A * A$ are equivalent to A. They are, however, *equi-inhabited*. This is a fancy way of saying

$$A \uplus A \text{ is inhabited} \quad \Leftrightarrow \quad A \text{ is inhabited}$$

The above is the real *essence* of the Curry–Howard correspondence. In other words, classical Curry–Howard tells us about *logical equivalence* of types. This is even a constructive statement: there are indeed functions $f : A \uplus A \to A$ and $g : A \to A \uplus A$; however, they are not inverses.

So mere inhabitation falls far short of our goals of being able to smoothly deform from one type to another. Let us thus analyze the crux of the "problem." In logic, we have that $\wedge$ and $\vee$ are both *idempotent*: this is the property of any binary operation $\circ$ where $\forall a.a \circ a = a$. And it should be clear that an idempotent operations is a *forgetful* operation: its input has two copies of a, but its output, only one. On the type side, something more subtle happens. Consider $\top \uplus \top$ versus $\top$; the first has exactly *two* proofs of inhabitation (left tt and right tt) while the second only one (tt). These cannot be put in bijective correspondence. Even though the "payload" tt is the same, forgetting left (or right) throws away information—something we have expressly disallowed. Yes, this should remind you of Maxwell's daemon: even though the data is the same, they are tagged differently, and these tags are indeed information, and their information content must be preserved.

Nevertheless, the Curry–Howard correspondence still has some force. We know that the inhabitants of types formed with $\perp$, $\top$, $\uplus$, $*$ form a commutative semiring. What we want to know is, which types are equivalent? From a commutative semiring perspective, this amounts to asking what terms are equal. We have a set of generators for those equations, namely those in Definition 2. What we thus need is to create 8 pairs of mutually inverse functions which witness these identities. For concreteness, we show the signatures in Fig. 1.

$$A \simeq A$$

$$
\begin{aligned}
\perp \uplus A &\simeq A \\
A \uplus B &\simeq B \uplus A \\
A \uplus (B \uplus C) &\simeq (A \uplus B) \uplus C
\end{aligned}
$$

$$
\begin{aligned}
\top * A &\simeq A \\
A * B &\simeq B * A \\
A * (B * C) &\simeq (A * B) * C
\end{aligned}
$$

$$
\begin{aligned}
\perp * A &\simeq \perp \\
(A \uplus B) * C &\simeq (A * C) \uplus (B * C)
\end{aligned}
$$

Fig. 1 Type isomorphisms.

From category theory, we are informed of the following privilege enjoyed by the natural numbers $\mathbb{N}$:

Theorem 1. *The semiring* $(\mathbb{N}, 0, 1, +, \cdot)$ *is* initial *in the category of semirings and semiring homomorphisms.*

In other words, for any semiring S, there is a homomorphism from $\mathbb{N}$ into S. But $\mathbb{N}$ is also the "counting" semiring, which formalizes the notion of cardinality of finite discrete sets.

The previous section on finite sets, along with the reasoning above, thus leads us to posit that the correct denotational semantics for finite discrete types is that of the semiring $(\mathbb{N}, 0, 1, +, \cdot)$. It is worth noting that equality in this semiring is intensional (i.e., two things are equal if and only if they are identical after evaluation), unlike that for types.

3.2 A language of type equivalences

We now have in our hands our desired denotational semantics for types. We want to create a programming language, which we call Π, such that the types and type combinators map to $\bot$, $\top$, $\uplus$, $*$, and such that we have ground terms whose denotation are all 16 type isomorphisms of Fig. 1. This is rather straightforward, as we can simply do this literally. To make the analogy with commutative semirings stand out even more, we will use 0, 1, +, and × at the type level, and will denote "equivalence" by $\leftrightarrow$. Thus Fig. 2 shows the "constants" of the language. As these all come in symmetric pairs (some of which are self-symmetric), we give names for both directions. Note how we have continued with the spirit of Curry–Howard: the terms of Π are *proof terms*, but rather than being witnesses of inhabitation, they are witnesses of equivalences. Thus we get an unexpected programming language design:

The proof terms denoting commutative semiring equivalences induce the terms of Π.

$$
\begin{array}{rrcll}
id_{\leftrightarrow}: & t & \leftrightarrow & t & : id_{\leftrightarrow} \\[6pt]
unite_{+}l: & 0 + t & \leftrightarrow & t & : uniti_{+}l \\
swap_{+}: & t_1 + t_2 & \leftrightarrow & t_2 + t_1 & : swap_{+} \\
assocl_{+}: & t_1 + (t_2 + t_3) & \leftrightarrow & (t_1 + t_2) + t_3 & : assocr_{+} \\[6pt]
unite_{\times}l: & 1 \times t & \leftrightarrow & t & : uniti_{\times}l \\
swap_{\times}: & t_1 \times t_2 & \leftrightarrow & t_2 \times t_1 & : swap_{\times} \\
assocl_{\times}: & t_1 \times (t_2 \times t_3) & \leftrightarrow & (t_1 \times t_2) \times t_3 & : assocr_{\times} \\[6pt]
absorbr: & 0 \times t & \leftrightarrow & 0 & : factorzl \\
dist: & (t_1 + t_2) \times t_3 & \leftrightarrow & (t_1 \times t_3) + (t_2 \times t_3) & : factor
\end{array}
$$

Fig. 2 Π-terms.

Of course, one does not get a programming language with just typed constants! There is a need to perform multiple equivalences. There are in fact three ways to do this: sequential composition $\odot$, choice composition $\oplus$ (sometimes called juxtaposition), and parallel composition $\otimes$. See Fig. 3 for the signatures. The construction $c_1 \odot c_2$ corresponds to performing c_1 first, then c_2, and is the usual notion of composition—and corresponds to ; of the language of permutations of Section 2. The construction $c_1 \oplus c_2$ chooses to perform c_1 or c_2 depending on whether the input is labeled left or right, respectively. Finally the construction $c_1 \otimes c_2$ operates on a product structure, and applies c_1 to the first component and c_2 to the second. The language of permutations lacked the ability to combine permutations by taking sums and products, which led to the awkward noncompositional programming style illustrated in the full adder example (Eq. 5).

Thus the denotation of the Π terms *should* be permutations. But given types A and B denoting $[m]$ and $[n]$ respectively, what are $A \uplus B$ and $A * B$? They correspond exactly to $[m+n]$ and $[m*n]$! Geometrically, this corresponds to concatenation for $A + B$, i.e., lining up the elements of A first, and then those of B. For $A * B$, one can picture this as lining up the elements of A horizontally, those of B vertically and perpendicular to those of A, and filling in the square with pairs of elements from A and B; if one renumbers these sequentially, reading row-wise, this gives an enumeration of $[m*n]$.

From here, it is easy to see what, for example, $c_1 \oplus c_2$ must be, operationally: from a permutation on $[m]$ and another on $[n]$, create a permutation on $[m+n]$ by having c_1 operate on the first m elements of $A + B$, and c_2 operate on the last n elements. Similarly, $swap_+$ switches the roles of A and B, and thus corresponds to $[n+m]$. Note how we "recover" the commutativity of natural number addition from this type isomorphism. Geometrically, $swap_\times$ is also rather interesting: it corresponds to matrix transpose! Furthermore, in this representations, some combinators like $unite_+l$ and $assocl_+$ are identity operations: the underlying representations are not merely isomorphic, they are definitionally equal. In other words, the passage to $\mathbb{N}$ erases some structural information.

$$\frac{\vdash c_1 : t_1 \leftrightarrow t_2 \quad \vdash c_2 : t_2 \leftrightarrow t_3}{\vdash c_1 \odot c_2 : t_1 \leftrightarrow t_3} \qquad \frac{\vdash c_1 : t_1 \leftrightarrow t_2 \quad \vdash c_2 : t_3 \leftrightarrow t_4}{\vdash c_1 \oplus c_2 : t_1 + t_3 \leftrightarrow t_2 + t_4}$$

$$\frac{\vdash c_1 : t_1 \leftrightarrow t_2 \quad \vdash c_2 : t_3 \leftrightarrow t_4}{\vdash c_1 \otimes c_2 : t_1 \times t_3 \leftrightarrow t_2 \times t_4}$$

Fig. 3 Π-combinators.

$$\frac{\vdash c_1 : t_1 \leftrightarrow t_2}{\vdash\ !\ c_1 : t_2 \leftrightarrow t_1}$$

Fig. 4 Derived Π-combinator.

$$
\begin{array}{llll}
unite_+r : & t + 0 & \leftrightarrow & t & : uniti_+r \\
unite_\times r : & t \times 1 & \leftrightarrow & t & : uniti_\times r \\
\\
absorbl : & t \times 0 & \leftrightarrow & 0 & : factorzr \\
distl : & t_1 \times (t_2 + t_3) & \leftrightarrow & (t_1 \times t_2) + (t_1 \times t_3) & : factorl
\end{array}
$$

Fig. 5 Additional Π-terms.

Embedded in our definition of Π is a conscious design decision: to make the terms of Π *syntactically* reversible. In other words, to every Π constant, there is another Π constant which is its inverse. As this is used frequently, we give it the short name !, and its type is given in Fig. 4. This combinator is *defined*, by pattern matching on the syntax of its argument and structural recursion.

This is not the only choice. Another would be to add a *flip* combinator to the language; we could then remove quite a few combinators as redundant. The drawback is that many programs in Π become longer. Furthermore, some of the symmetry at "higher levels" (see next section) is also lost. Since the extra burden of language definition and of proofs is quite low, we prefer the structural symmetry over a minimalistic language definition.

We also make a second design decision, which is to make the Π language itself symmetric in another sense: we want both left and right introduction/ elimination rules for units, 0 absorption and distributivity. Specifically, we add the Π-terms of Fig. 5 to our language. These are redundant because of $swap_+$ and $swap_\times$, but will later enable shorter programs and more elegant presentation of program transformations.

This set of isomorphisms is known to be sound and complete [39, 40] for isomorphisms of finite types. Furthermore, it is also universal for hardware combinational circuits [31].

3.3 Operational semantics

To give an operational semantics to Π, we are mainly missing a notation for *values*.

Definition 8. (Syntax of values of Π)

$$values, v \quad ::= \quad () \mid left\ v \mid right\ v \mid (v, v)$$

Given a program $c : b_1 \leftrightarrow b_2$ in Π, we can run it by supplying it with a value $v_1 : b_1$. The evaluation rules $c\, v_1 \mapsto v_2$ are given below.

Definition 9. (Operational Semantics for Π)

Identity:

$$id\!\leftrightarrow \quad v \quad \mapsto \quad v$$

Additive fragment:

$$
\begin{array}{llll}
unite_+l & (right\ v) & \mapsto & v \\
uniti_+l & v & \mapsto & right\ v \\
unite_+r & (left\ v) & \mapsto & v \\
uniti_+r & v & \mapsto & left\ v \\
swap_+ & (left\ v) & \mapsto & right\ v \\
swap_+ & (right\ v) & \mapsto & left\ v \\
assocl_+ & (left\ v_1) & \mapsto & left\ (left\ v_1) \\
assocl_+ & (right\ (left\ v_2)) & \mapsto & left\ (right\ v_2) \\
assocl_+ & (right\ (right\ v_3)) & \mapsto & right\ v_3 \\
assocr_+ & (left\ (left\ v_1)) & \mapsto & left\ v_1 \\
assocr_+ & (left\ (right\ v_2)) & \mapsto & right\ (left\ v_2) \\
assocr_+ & (right\ v_3) & \mapsto & right\ (right\ v_3)
\end{array}
$$

Multiplicative fragment:

$$
\begin{array}{llll}
unite_\times l & ((),v) & \mapsto & v \\
uniti_\times l & v & \mapsto & ((),v) \\
unite_\times r & (v,()) & \mapsto & v \\
uniti_\times r & v & \mapsto & (v,()) \\
swap_\times & (v_1,v_2) & \mapsto & (v_2,v_1) \\
assocl_\times & (v_1,(v_2,v_3)) & \mapsto & ((v_1,v_2),v_3) \\
assocr_\times & ((v_1,v_2),v_3) & \mapsto & (v_1,(v_2,v_3)) \\
absorbr & (v_1,v_2) & \mapsto & v_1
\end{array}
$$

Distributivity and factoring:

$$
\begin{array}{rcll}
dist & (left\ v_1, v_3) & \mapsto & left\ (v_1, v_3) \\
dist & (right\ v_2, v_3) & \mapsto & right\ (v_2, v_3) \\
distl & (v_1, left\ v_2) & \mapsto & left\ (v_1, v_2) \\
distl & (v_1, right\ v_3) & \mapsto & right\ (v_1, v_3) \\
factor & (left\ (v_1, v_3)) & \mapsto & (left\ v_1, v_3) \\
factor & (right\ (v_2, v_3)) & \mapsto & (right\ v_2, v_3) \\
factorl & (left\ (v_1, v_2)) & \mapsto & (v_1, left\ v_2) \\
factorl & (right\ (v_1, v_3)) & \mapsto & (v_1, right\ v_3) \\
absorbl & (v_1, v_2) & \mapsto & v_2
\end{array}
$$

The evaluation rules of the composition combinators are given below:

$$
\frac{c_1\ v_1 \mapsto v \qquad c_2\ v \mapsto v_2}{(c_1 \odot c_2)\ v_1 \mapsto v_2}
$$

$$
\frac{c_1\ v_1 \mapsto v_2}{(c_1 \oplus c_2)\ (left\ v_1) \mapsto left\ v_2}
\qquad
\frac{c_2\ v_1 \mapsto v_2}{(c_1 \oplus c_2)\ (right\ v_1) \mapsto right\ v_2}
$$

$$
\frac{c_1\ v_1 \mapsto v_3 \qquad c_2\ v_2 \mapsto v_4}{(c_1 \otimes c_2)\ (v_1, v_2) \mapsto (v_3, v_4)}
$$

Since there are no values that have the type 0, the reductions for the combinators $unite_+l$, $uniti_+l$, $unite_+r$, and $uniti_+r$ omit the impossible cases. $factorzr$ and $factorzl$ likewise do not appear as they have no possible cases at all. However, $absorbr$ and $absorbl$ are treated slightly differently: rather than *eagerly* assuming they are impossible, the purported inhabitant of 0 given on one side is passed on to the other side. The reason for this choice will have to wait for Section 4.2 when we explain some higher level symmetries (see Fig. 13).

As we mentioned before, ! is a defined combinator.

Definition 10 (Adjoint, ! c). The adjoint of a combinator c is defined as follows:

- For primitive isomorphisms c, ! c is given by its inverse from Figs. 2 and 5.
- $!(c_1 \otimes c_2) = !c_1 \otimes !c_2$
- $!(c_1 \oplus c_2) = !c_1 \oplus !c_2$
- $!(c_1 \odot c_2) = !c_2 \odot !c_1$. (Note that the order of combinators has been reversed).

We can further define that two combinators are *observationally equivalent* if on all values of their common domain, they evaluate to identical values. More precisely, we will say that for combinators $c_1, c_2 : b_1 \leftrightarrow b_2$, $c_1 = c_2$ whenever:

$$\forall\; v_1 : b_1, v_2 : b_2.\; c_1\; v_1 \mapsto v_2 \text{ if and only if } c_2\; v_1 \mapsto v_2$$

Each type b has a size $|b|$ defined in the obvious way. We had previously established that for any natural number n, there is a canonical set of size n, which we denoted $[n]$. Furthermore, we can also define a canonical *type* of that size, which we will denote $\sharp\, b$, i.e., $\sharp\, b$ is a canonical type of size $|b|$.

Definition 11. ($\sharp$). By recursion on $|b|$. First define τ that maps numeric sizes to their corresponding types. We will revert to using type notation for greater clarity of this definition:

$$
\begin{aligned}
\tau\,(0) &= \bot \\
\tau\,(1+n) &= \top \uplus \tau\,(n)
\end{aligned}
$$

so that we can define $\sharp\, b = \tau\,|b|$.

We are now ready to go further and establish that there is always an equivalence between a type and the canonical type of the same size.

Proposition 2. *For any type b there exists an isomorphism $b \leftrightarrow \sharp\, b$.*

Proof. The fact that such an isomorphism exists is evident from the definition of size and what it means for two types to be isomorphic. While many equivalent constructions are possible for any type b, one such construction is given by $[\![b]\!]$:

$$
\begin{aligned}
[\![0]\!] &= id \leftrightarrow \\
[\![1]\!] &= id \leftrightarrow \\
[\![1+b]\!] &= id \leftrightarrow \oplus [\![b]\!] \\
[\![(b_1+b_2)+b_3]\!] &= assocr_+ \odot [\![b_1 + (b_2 + b_3)]\!] \\
[\![b_1+b_2]\!] &= ([\![b_1]\!] \oplus id \leftrightarrow) \odot [\![\#b_1 + b_2]\!] \\
[\![0 \times b_2]\!] &= absorbr \\
[\![1 \times b_2]\!] &= unite_\times l \odot [\![b_2]\!] \\
[\![(b_1 \times b_2) \times b_3]\!] &= assocr_\times \odot [\![b_1 \times (b_2 \times b_3)]\!] \\
[\![(b_1+b_2) \times b_3]\!] &= dist \odot [\![b_1 \times b_3 + b_2 \times b_3]\!]
\end{aligned}
$$

$\square$

3.4 Graphical language

Combinators of Π can be written in terms of the operators described previously or via a graphical language similar in spirit to those developed for Geometry of Interaction [41] and string diagrams for category theory [42, 43]. Modulo some conventions and shorthand we describe here, the wiring diagrams are equivalent to the operator based (syntactic) description of programs. Π combinators expressed in this graphical language look like "wiring diagrams." Values take the form of "particles" that flow along the wires. Computation is expressed by the flow of particles.

- The simplest sort of diagram is the $id{\leftrightarrow}: b \leftrightarrow b$ combinator which is simply represented as a wire labeled by its type b. In more complex diagrams, if the type of a wire is obvious from the context, it may be omitted.

$$b$$

Values flow from left to right in the graphical language of Π. When tracing a computation, one might imagine a value v of type b on the wire, as shown below.

- The product type $b_1 \times b_2$ may be represented both as one wire labeled $b_1 \times b_2$ or by two parallel wires labeled b_1 and b_2. Both representations may be used interchangeably.

$$b_1 \times b_2 \qquad\qquad \begin{array}{c} b_1 \\ \\ b_1 \times b_2 \\ \\ b_2 \end{array}$$

When tracing execution using particles, one should think of one particle on each wire or alternatively as in folklore in the literature on monoidal categories as a "wave."

- Sum types may similarly be represented using parallel wires with a + operator between them.

When tracing the execution of $b_1 + b_2$ represented by one wire, one can think of a value of the form *left* v_1 or *right* v_2 as flowing on the wire, where $v_1 : b_1$ and $v_2 : b_2$. When tracing the execution of two additive wires, a value can reside on only one of the two wires.

$$\underset{}{\overset{v_1 : b_1}{\rule{3cm}{0.4pt}}}$$

- When representing complex types like $(b_1 \times b_2) + b_3$ some visual grouping of the wires may be done to aid readability. The exact type however will always be clarified by the context of the diagram.

$$\begin{array}{c} \underline{\quad b_1 \quad} \\ \underline{\quad b_2 \quad} \\ + \\ \underline{\quad b_3 \quad} \end{array}$$

- Associativity is entirely skipped in the graphical language. Hence three parallel wires may be inferred as $b_1 \times (b_2 \times b_3)$ or $(b_1 \times b_2) \times b_3$, based on the context. This is much like handling of associativity in the graphical representations of categories as well as that for monoidal categories.

$$\begin{array}{c} \underline{\quad b_1 \quad} \\ \underline{\quad b_2 \quad} \\ \underline{\quad b_3 \quad} \end{array}$$

- Commutativity is represented by crisscrossing wires.

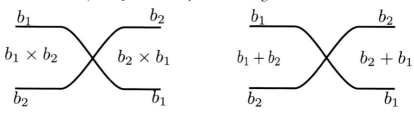

When tracing the execution of b_1+b_2 represented by one wire, one can think of a value of the form *left v_1* or *right v_2* as flowing on the wire, where $v_1 : b_1$ and $v_2 : b_2$. By visually tracking the flow of particles on the wires, one can verify that the expected types for commutativity are satisfied.

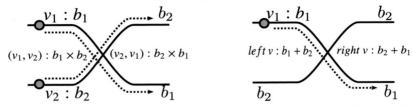

- The morphisms that witness that 0 and 1 are the additive and multiplicative units are represented as shown below. Note that since there is no value of type 0, there can be no particle on a wire of type 0. Also since the monoidal units can be freely introduced and eliminated, sometimes they are omitted. However, as this is in fact dangerous, as explained by [42], we will err on the side of including them.

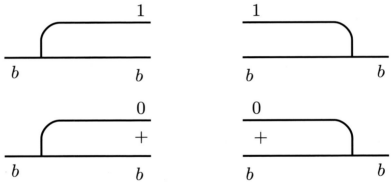

- Distributivity and factoring are represented using the dual boxes shown below:

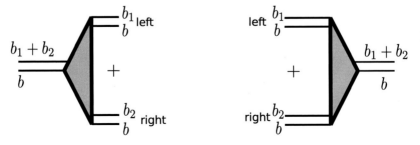

Distributivity and factoring are interesting because they represent interactions between sum and pair types. Distributivity should essentially be thought of as a multiplexer that redirects the flow of $v : b$ depending on what value inhabits the type $b_1 + b_2$, as shown below.

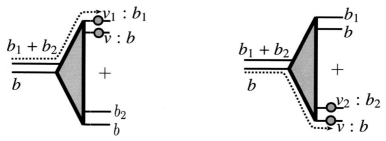

Factoring is the corresponding adjoint operation.

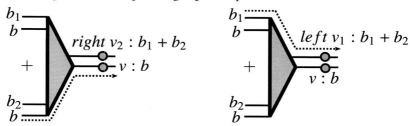

- Combinators can be composed in series ($c_1 \odot c_2$) or parallel. Sequential (series) composition corresponds to connecting the output of one combinator to the input of the next.

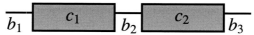

There are two forms of parallel composition—combinators can be combined additively $c_1 \oplus c_2$ (shown on the left) or multiplicatively $c_1 \otimes c_2$ (shown on the right).

Example. As an example consider the wiring diagram of the combinator c below:

$$c \quad : \quad b \times (1 + 1) \leftrightarrow b + b$$
$$c \quad = \quad swap_\times \odot dist \odot (unite_\times l \oplus unite_\times l)$$

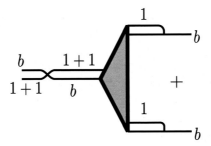

3.5 Denotational semantics

Fig. 1 introduces our desired denotational semantics, and Section 3.3 is a direct definition of an operational semantics. One obvious question arises: do these correspond?

We can certainly associate to each Π combinator an equivalence between the denotation of each type[b]:

$$\mathsf{c2equiv} : \{t_1 \ t_2 : \mathsf{U}\} \to (c : t_1 \leftrightarrow t_2) \to [\![\, t_1 \,]\!] \simeq [\![\, t_2 \,]\!]$$

And as such an equivalence contains a function as its first component, we can compare if our operational semantics and denotational semantics match. And they do:

$$\mathsf{lemma0} : \{t_1 \ t_2 : \mathsf{U}\} \to (c : t_1 \leftrightarrow t_2) \to (v : [\![\, t_1 \,]\!]) \to \mathsf{eval} \ c \ v \equiv \mathsf{proj}_1 \ (\mathsf{c2equiv} \ c) \ v$$

We can similarly hand-write a backwards evaluator, prove that it is indeed a proper backwards evaluator, and finally show that it agrees with the reverse equivalence.

3.6 Examples

At first, it is not immediately clear that a programming language in which information is preserved could model choice. We recall a quote by Minsky communicating this concern:

> Ed Fredkin pursued the idea that information must be finite in density. One day, he announced that things must be even more simple than that. He said that he was going to assume that information itself is conserved. "You're out of you mind, Ed." I pronounced. "That's completely ridiculous. Nothing could happen in such a world. There couldn't even be logical gates. No decisions could ever be made." But when Fredkin gets one of his ideas, he's quite immune to objections like that; indeed, they fuel him with energy. Soon he went on to assume that information processing must also be reversible—and invented what's now called the Fredkin gate [33].

---

[b] This is extracted from the Agda formalization of this work, which has been reported on in a previous paper [32].

We will however show that one can program all logical gates in Π. We will start with a few simple examples and then discuss the expressiveness of the language and its properties.

3.6.1 Booleans

Let us start with encoding Booleans. We use the type $1+1$ to represent Booleans with *left* () representing *true* and *right* () representing *false*. Boolean negation is straightforward to define:

$$not : bool \leftrightarrow bool$$

$$not = swap_+$$

It is easy to verify that *not* changes *true* to *false* and vice versa.

3.6.2 Bit vectors

We can represent n-bit words using an n-ary product of *bools*. For example, we can represent a 3-bit word, $word_3$, using the type $bool \times (bool \times bool)$. We can perform various operations on these 3-bit words using combinators in Π. For instance the bitwise *not* operation is the parallel composition of three *not* operations:

$$not_{word_3} : word_3 \leftrightarrow word_3$$

$$not_{word_3} = not \times (not \times not)$$

We can express a 3-bit word reversal operation as follows:

$$reverse : word_3 \leftrightarrow word_3$$

$$reverse = swap_\times \odot (swap_\times \otimes id\leftrightarrow) \odot assocr_\times$$

We can check that *reverse* does the right thing by applying it to a value $(v_1, (v_2, v_3))$ and writing out the full derivation tree of the reduction. The combinator *reverse*, like many others we will see in this paper, is formed by sequentially composing several simpler combinators. Instead of presenting the operation of *reverse* as a derivation tree, it is easier (purely for presentation reasons) to flatten the tree into a sequence of reductions as caused by each component. Such a sequence of reductions is given below:

$$(v_1, (v_2, v_3))$$

$$swap_\times \quad ((v_2, v_3), v_1)$$

$$swap_\times \otimes id\leftrightarrow \quad ((v_3, v_2), v_1)$$

$$assocr_\times \quad (v_3, (v_2, v_1))$$

On the first line is the initial value. On each subsequent line is a fragment of the *reverse* combinator and the value that results from applying this combinator to the value on the previous line. For example, $swap_\times$ transforms $(v_1, (v_2, v_3))$ to $((v_2, v_3), v_1)$. On the last line we see the expected result with the bits in reverse order.

We can also draw out the graphical representation of the 3-bit reverse combinator. In the graphical representation, it is clear that the combinator achieves the required shuffling.

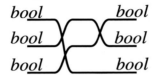

3.6.3 Conditionals

Even though Π lacks conditional expressions, they are expressible using the distributivity and factoring laws. The diagrammatic representation of *dist* shows that it redirects the flow of a value $v : b$ based on the value of another one of type $b_1 + b_2$. If we choose $1 + 1$ to be *bool* and apply either $c_1 : b_1 \leftrightarrow b_2$ or $c_2 : b_1 \leftrightarrow b_2$ to the value v, then we essentially have an 'if' expression.

$$if_{c_1,c_2} : bool \times b_1 \leftrightarrow bool \times b_2$$
$$if_{c_1,c_2} = dist \odot \left((id \leftrightarrow \otimes c_1) + (id \leftrightarrow \otimes c_2)\right) \odot factor$$

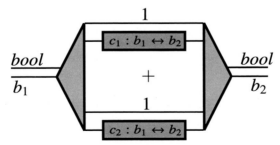

The diagram above shows the input value of type $(1+1) \times b_1$ processed by the distribute operator *dist*, which converts it into a value of type $(1 \times b_1) + (1 \times b_1)$. In the *left* branch, which corresponds to the case when the Boolean is *true* (i.e., the value was *left* ()), the combinator c_1 is applied to the value of type b_1. The right branch which corresponds to the Boolean being *false* passes the value of type b_1 through the combinator c_2. The inverse of *dist*, namely *factor* is applied to get the final result of type $(1+1) \times b_2$.

3.6.4 Logic gates

There are several universal primitives for conventional (irreversible) hardware circuits, such as *nand* and *fanout*. In the case of reversible hardware circuits, the canonical universal primitive is the Toffoli gate [1]. The Toffoli gate takes three Boolean inputs: if the first two inputs are *true* then the third bit is negated. In a traditional language, the Toffoli gate would be most conveniently expressed as a conditional expression like:

$$toffoli(v_1, v_2, v_3) = if \ (v_1 \ and \ v_2) \ then \ (v_1, v_2, not(v_3)) \ else \ (v_1, v_2, v_3)$$

We will derive Toffoli gate in Π by first deriving a simpler logic gate called *cnot*. Consider a one-armed version, if_c, of the conditional derived above. If the *bool* is *true*, the value of type b is modified by the operator c.

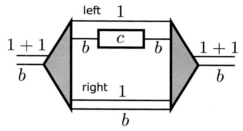

By choosing b to be *bool* and c to be *not*, we have the combinator $if_{not} : bool \times bool \leftrightarrow bool \times bool$ which negates its second argument if the first argument is *true*. This gate if_{not} is often referred to as the *cnot* gate[1].

If we iterate this construction once more, the resulting combinator if_{cnot} has type $bool \times (bool \times bool) \leftrightarrow bool \times (bool \times bool)$. The resulting gate checks the first argument and if it is *true*, proceeds to check the second argument. If that is also *true* then it will negate the third argument. Thus if_{cnot} is the required Toffoli gate.

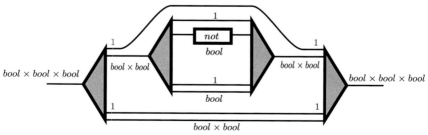

4. Data III: Reversible programs between reversible programs

In Sections 2 and 3, we examined equivalences between conventional data structures, i.e., sets of values and structured trees of values. We now consider a richer but foundational notion of data: programs themselves. Indeed, universal computation models crucially rely on the fact that *programs are (or can be encoded as) data*, e.g., a Turing Machine can be encoded as a string that another Turing Machine (or even the same machine) can manipulate. Similarly, first-class functions are the *only* values in the λ-calculus. In our setting, the programs developed in the previous section are reversible deformations between structured finite types. We now ask whether these programs can themselves be subject to (higher level) reversible deformations?

Before developing the theory, let's consider a small example consisting of two deformations between the types $A + B$ and $C + D$:

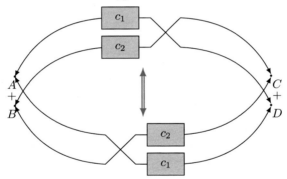

The top path is the Π program $(c_1 \oplus c_2) \odot swap_+$ which deforms the type A by c_1, deforms the type B by c_2, and deforms the resulting space by a twist that exchanges the two injections into the sum type. The bottom path performs the twist first and then deforms the type A by c_1 and the type B by c_2 as before. One could imagine the paths are physical *elastic* wires in 3 space, where the deformations c_1 and c_2 as arbitrary deformations on these wires, and the twists do not touch but are in fact well-separated. Then, holding the points A, B, C, and D fixed, it is possible to imagine sliding c_1 and c_2 from the top wire rightward past the twist, and then using the elasticity of the wires, pull the twist back to line up with that of the bottom—thus making both parts of the diagram identical. Each of these moves can be undone (reversed), and doing so would take the bottom part of the diagram into the top part. In other words, there exists a deformation of the program $(c_1 \oplus c_2) \odot swap_+$ to the program $swap_+ \odot (c_2 \oplus c_1)$. We can also show that this means that, as permutations, $(c_1 \oplus c_2) \odot swap_+$ and $swap_+ \odot (c_2 \oplus c_1)$ are equal. And, of course, not all programs between the same types can be deformed into one another. The simplest example of inequivalent deformations is the two automorphisms of $1 + 1$, namely $id \leftrightarrow$ and $swap_+$.

While we will not make the details of the stretchable wires and slidable boxes formal, it is useful for intuition. One caveat though: some of the sliding and stretching needs to be done in spaces of higher dimension than 3 to have "enough room" to move things along without collision or over-stretching wires. That, unfortunately, means that some equivalences are harder to grasp. Luckily, most equivalences only need 3 dimensions.

Our reversible language of type isomorphisms and equivalences between them has a strong connection to *univalent universes* in HoTT [44]. Based on this connection, we refer to the types as being at level-0, to the equivalences between types (i.e., the combinators of Section 3) as being at level-1, and to the equivalences between equivalences of types (i.e., the combinators discussed in this section) as being at level-2.

4.1 A model of equivalences between type equivalences

Previously we saw how we could take the proof terms of commutative semiring equivalences as our starting point for Π. What we need now is to understand how *proofs* of algebraic identities should be considered equivalent. Classical algebra does not help, as proofs are not considered first-class citizens. However, another route is available to us: since the work of Hofmann and Streicher [45], we know that one can model types as *groupoids*. The additional structure comes from explicitly modeling the "identity types": instead of regarding all terms which witness the equality of (say) a and b of type A as being indistinguishable, we posit that there may in fact be many. The consequences of this one decision are enough to show that types can be modeled by groupoids.

Thus, rather than looking at (untyped) commutative semirings, we should look at a *typed* version. This process frequently goes by the moniker of "categorification." We want a categorical algebra, where the basic objects are groupoids (to model our types), and where there is a natural notion of $+$ and $*$. At first, we hit what seems like a serious stumbling block: the category of all groupoids, **Groupoid**, have neither coproducts nor products. However, we don't want to work internally in **Groupoid**—we want operations *on* groupoids. In other words, we want something akin to symmetric monoidal categories, but with two interacting monoidal structures. Luckily, this already exists: the categorical analog to (commutative) semirings are (symmetric) Rig Categories [46, 47]. This straightforwardly generalizes to symmetric Rig Groupoids.

How does this help? Coherence conditions! Symmetric monoidal categories, to start somewhere simple, do not just introduce natural transformations like the associator α and the left and right unitors (λ and ρ respectively), but also coherence conditions that these must satisfy. Looking, for example, at just the additive fragment of Π (i.e., with just 0, 1 and + for the types, $\odot$ and $\oplus$ as combinators, and only the terms so expressible), the sublanguage would correspond, denotationally, to exactly (nonempty) symmetric monoidal groupoids. And what these possess are exactly some *equations between equations* as commutative diagrams. Transporting these coherence conditions, for example those that express that various transformations are *natural* to Π, gives a list of equations between Π programs. Furthermore, all the natural transformations that arise are in fact natural *isomorphisms*—and thus reversible.

We can then proceed to prove that every one of the coherence conditions involved in defining a symmetric Rig Groupoid holds for the groupoid interpretation of types [32]. This is somewhat tedious given the sheer number of these, but when properly formulated, relatively straightforward, but see below for comments on some tricky cases.

But why are these particular coherence laws? Are they all necessary? Conversely are they, in some appropriate sense, sufficient? This is the so-called *coherence problem*. Mac Lane, in his farewell address as President of the American Mathematical Society [48] gives a good introduction and overview of such problems. A more modern interpretation (which can nevertheless be read into Mac Lane's own exposition) would read as follows: given a set of equalities on abstract words, regarded as a rewrite system, and two means of rewriting a word in that language to another, is there some suitable notion of canonical form that expresses the essential uniqueness of the nontrivial rewrites? Note how this word-and-rewrite problem is essentially independent of the eventual interpretation. But one must take some care, as there are obvious degenerate cases (involving "trivial" equations involving 0 or 1) which lead to nonuniqueness. The landmark results, first by Kelly-Mac Lane [49] for closed symmetric monoidal categories, then (independently) Laplaza and Kelly [46, 47] for symmetric Rig Categories, is that indeed there are sound and complete coherence conditions that insure that all the "obvious" equalities between different abstract words in these systems give rise to commutative diagrams. The "obvious" equalities come from *syzygies* or *critical pairs* of the system of equations. The problem is far from trivial—Fiore et al. [50] document some publications where the coherence set is in fact incorrect. They furthermore give a quite general algorithm to derive such coherence conditions.

4.2 A language of equivalences between type equivalences

As motivated in the previous section, the equivalences between type equivalences are perfectly modeled by the coherence conditions of weak Rig Groupoids. Syntactically, we take the easiest way there: simply make every coherence isomorphism into a programming construct. These constructs are collected in several figures (Figs. 7–15) and are discussed next.

Conveniently, the various coherence conditions can be naturally grouped into "related" laws. Each group basically captures the interactions between compositions of level-1 Π combinators.

Starting with the simplest constructions, the first two constructs in Fig. 6 are the level-2 analogs of $+$ and $*$, which respectively model level-1 choice composition and parallel composition (of equivalences). These allow us to "build up" larger equivalences from smaller ones. The next two express that

Let $c_1 : t_1 \leftrightarrow t_2$, $c_2 : t_3 \leftrightarrow t_4$, $c_3 : t_1 \leftrightarrow t_2$, and $c_4 : t_3 \leftrightarrow t_4$.
Let $a_1 : t_5 \leftrightarrow t_1$, $a_2 : t_6 \leftrightarrow t_2$, $a_3 : t_1 \leftrightarrow t_3$, and $a_4 : t_2 \leftrightarrow t_4$.

$$\frac{c_1 \Leftrightarrow c_3 \quad c_2 \Leftrightarrow c_4}{c_1 \oplus c_2 \Leftrightarrow c_3 \oplus c_4} \qquad \frac{c_1 \Leftrightarrow c_3 \quad c_2 \Leftrightarrow c_4}{c_1 \otimes c_2 \Leftrightarrow c_3 \otimes c_4}$$

$$(a_1 \odot a_3) \oplus (a_2 \odot a_4) \Leftrightarrow (a_1 \oplus a_2) \odot (a_3 \oplus a_4)$$

$$(a_1 \odot a_3) \otimes (a_2 \odot a_4) \Leftrightarrow (a_1 \otimes a_2) \odot (a_3 \otimes a_4)$$

Fig. 6 Signatures of level-2 Π-combinators: functors.

Let $c_1 : t_1 \leftrightarrow t_2$, $c_2 : t_2 \leftrightarrow t_3$, and $c_3 : t_3 \leftrightarrow t_4$:

$$c_1 \odot (c_2 \odot c_3) \Leftrightarrow (c_1 \odot c_2) \odot c_3$$

$$(c_1 \oplus (c_2 \oplus c_3)) \odot assocl_+ \Leftrightarrow assocl_+ \odot ((c_1 \oplus c_2) \oplus c_3)$$

$$(c_1 \otimes (c_2 \otimes c_3)) \odot assocl_\times \Leftrightarrow assocl_\times \odot ((c_1 \otimes c_2) \otimes c_3)$$

$$((c_1 \oplus c_2) \oplus c_3) \odot assocr_+ \Leftrightarrow assocr_+ \odot (c_1 \oplus (c_2 \oplus c_3))$$

$$((c_1 \otimes c_2) \otimes c_3) \odot assocr_\times \Leftrightarrow assocr_\times \odot (c_1 \otimes (c_2 \otimes c_3))$$

$$assocr_+ \odot assocr_+ \Leftrightarrow ((assocr_+ \oplus id\leftrightarrow) \odot assocr_+) \odot (id\leftrightarrow \oplus assocr_+)$$

$$assocr_\times \odot assocr_\times \Leftrightarrow ((assocr_\times \otimes id\leftrightarrow) \odot assocr_\times) \odot (id\leftrightarrow \otimes assocr_\times)$$

Fig. 7 Signatures of level-2 Π-combinators: associativity.

Let $c_1 : t_1 \leftrightarrow t_2$, $c_2 : t_3 \leftrightarrow t_4$, and $c_3 : t_5 \leftrightarrow t_6$:

$$((c_1 \oplus c_2) \otimes c_3) \odot dist \Leftrightarrow dist \odot ((c_1 \otimes c_3) \oplus (c_2 \otimes c_3))$$
$$(c_1 \otimes (c_2 \oplus c_3)) \odot distl \Leftrightarrow distl \odot ((c_1 \otimes c_2) \oplus (c_1 \otimes c_3))$$
$$((c_1 \otimes c_3) \oplus (c_2 \otimes c_3)) \odot factor \Leftrightarrow factor \odot ((c_1 \oplus c_2) \otimes c_3)$$
$$((c_1 \otimes c_2) \oplus (c_1 \otimes c_3)) \odot factorl \Leftrightarrow factorl \odot (c_1 \otimes (c_2 \oplus c_3))$$

Fig. 8 Signatures of level-2 Π-combinators: distributivity and factoring.

Let $c_0, c_1, c_2, c_3 : t_1 \leftrightarrow t_2$ and $c_4, c_5 : t_3 \leftrightarrow t_4$:

$$id{\leftrightarrow} \odot c_0 \Leftrightarrow c_0 \quad c_0 \odot id{\leftrightarrow} \Leftrightarrow c_0 \quad c_0 \odot {!}c_0 \Leftrightarrow id{\leftrightarrow} \quad {!}c_0 \odot c_0 \Leftrightarrow id{\leftrightarrow}$$
$$id{\leftrightarrow} \oplus id{\leftrightarrow} \Leftrightarrow id{\leftrightarrow} \qquad id{\leftrightarrow} \otimes id{\leftrightarrow} \Leftrightarrow id{\leftrightarrow}$$

$$c_0 \Leftrightarrow c_0 \qquad \frac{c_1 \Leftrightarrow c_2 \quad c_2 \Leftrightarrow c_3}{c_1 \Leftrightarrow c_3} \qquad \frac{c_1 \Leftrightarrow c_4 \quad c_2 \Leftrightarrow c_5}{c_1 \odot c_2 \Leftrightarrow c_4 \odot c_5}$$

Fig. 9 Signatures of level-2 Π-combinators: identity and composition.

Let $c_0 : 0 \leftrightarrow 0$, $c_1 : 1 \leftrightarrow 1$, and $c_3 : t_1 \leftrightarrow t_2$:

$$unite_+l \odot c_3 \Leftrightarrow (c_0 \oplus c_3) \odot unite_+l \qquad uniti_+l \odot (c_0 \oplus c_3) \Leftrightarrow c_3 \odot uniti_+l$$
$$unite_+r \odot c_3 \Leftrightarrow (c_3 \oplus c_0) \odot unite_+r \qquad uniti_+r \odot (c_3 \oplus c_0) \Leftrightarrow c_3 \odot uniti_+r$$
$$unite_\times l \odot c_3 \Leftrightarrow (c_1 \otimes c_3) \odot unite_\times l \qquad uniti_\times l \odot (c_1 \otimes c_3) \Leftrightarrow c_3 \odot uniti_+l$$
$$unite_\times r \odot c_3 \Leftrightarrow (c_3 \otimes c_1) \odot unite_\times r \qquad uniti_\times r \odot (c_3 \otimes c_1) \Leftrightarrow c_3 \odot uniti_\times r$$
$$unite_\times l \Leftrightarrow distl \odot (unite_\times l \oplus unite_\times l)$$
$$unite_+l \Leftrightarrow swap_+ \odot unite_+r \qquad unite_\times l \Leftrightarrow swap_\times \odot unite_\times r$$

Fig. 10 Signatures of level-2 Π-combinators: unit.

Let $c_1 : t_1 \leftrightarrow t_2$ and $c_2 : t_3 \leftrightarrow t_4$:

$$swap_+ \odot (c_1 \oplus c_2) \Leftrightarrow (c_2 \oplus c_1) \odot swap_+ \quad swap_\times \odot (c_1 \otimes c_2) \Leftrightarrow (c_2 \otimes c_1) \odot swap_\times$$
$$(assocr_+ \odot swap_+) \odot assocr_+ \Leftrightarrow ((swap_+ \oplus id{\leftrightarrow}) \odot assocr_+) \odot (id{\leftrightarrow} \oplus swap_+)$$
$$(assocl_+ \odot swap_+) \odot assocl_+ \Leftrightarrow ((id{\leftrightarrow} \oplus swap_+) \odot assocl_+) \odot (swap_+ \oplus id{\leftrightarrow})$$
$$(assocr_\times \odot swap_\times) \odot assocr_\times \Leftrightarrow ((swap_\times \otimes id{\leftrightarrow}) \odot assocr_\times) \odot (id{\leftrightarrow} \otimes swap_\times)$$
$$(assocl_\times \odot swap_\times) \odot assocl_\times \Leftrightarrow ((id{\leftrightarrow} \otimes swap_\times) \odot assocl_\times) \odot (swap_\times \otimes id{\leftrightarrow})$$

Fig. 11 Signatures of level-2 Π-combinators: commutativity and associativity.

$$unite_+r \oplus id{\leftrightarrow} \quad \Leftrightarrow \quad assocr_+ \odot (id{\leftrightarrow} \oplus unite_+l)$$
$$unite_\times r \otimes id{\leftrightarrow} \quad \Leftrightarrow \quad assocr_\times \odot (id{\leftrightarrow} \otimes unite_\times l)$$

Fig. 12 Signatures of level-2 Π-combinators: unit and associativity.

Let $c : t_1 \leftrightarrow t_2$:

$$(c \otimes id\leftrightarrow) \odot absorbl \Leftrightarrow absorbl \odot id\leftrightarrow \quad (id\leftrightarrow \otimes c) \odot absorbr \Leftrightarrow absorbr \odot id\leftrightarrow$$

$$id\leftrightarrow \odot factorzl \Leftrightarrow factorzl \odot (id\leftrightarrow \otimes c) \quad id\leftrightarrow \odot factorzr \Leftrightarrow factorzr \odot (c \otimes id\leftrightarrow)$$

$$absorbr \Leftrightarrow absorbl$$

$$absorbr \Leftrightarrow (distl \odot (absorbr \oplus absorbr)) \odot unite_{+}l$$

$$unite_{\times} r \Leftrightarrow absorbr \qquad absorbl \Leftrightarrow swap_{\times} \odot absorbr$$

$$absorbr \Leftrightarrow (assocl_{\times} \odot (absorbr \otimes id\leftrightarrow)) \odot absorbr$$

$$(id\leftrightarrow \otimes absorbr) \odot absorbl \Leftrightarrow (assocl_{\times} \odot (absorbl \otimes id\leftrightarrow)) \odot absorbr$$

$$id\leftrightarrow \otimes unite_{+}l \Leftrightarrow (distl \odot (absorbl \oplus id\leftrightarrow)) \odot unite_{+}l$$

Fig. 13 Signatures of level-2 Π-combinators: zero.

$$((assocl_{+} \otimes id\leftrightarrow) \odot dist) \odot (dist \oplus id\leftrightarrow) \Leftrightarrow (dist \odot (id\leftrightarrow \oplus dist)) \odot assocl_{+}$$

$$assocl_{\times} \odot distl \Leftrightarrow ((id\leftrightarrow \otimes distl) \odot distl) \odot (assocl_{\times} \oplus assocl_{\times})$$

$$(distl \odot (dist \oplus dist)) \odot assocl_{+} \quad \Leftrightarrow \quad dist \odot (distl \oplus distl) \odot assocl_{+} \odot$$
$$(assocr_{+} \oplus id\leftrightarrow) \odot$$
$$((id\leftrightarrow \oplus swap_{+}) \oplus id\leftrightarrow) \odot$$
$$(assocl_{+} \oplus id\leftrightarrow)$$

Fig. 14 Signatures of level-2 Π-combinators: associativity and distributivity.

$$(id\leftrightarrow \otimes swap_{+}) \odot distl \quad \Leftrightarrow \quad distl \odot swap_{+}$$

$$dist \odot (swap_{\times} \oplus swap_{\times}) \quad \Leftrightarrow \quad swap_{\times} \odot distl$$

Fig. 15 Signatures of level-2 Π-combinators: commutativity and distributivity.

both of these composition operators distribute over sequential composition $\odot$ (and vice versa).

The constructs in Fig. 7 capture the informal idea that all the different ways of associating programs are equivalent. The first says that sequential composition itself ($\odot$) is associative. The next four capture how the $\oplus$ and $\otimes$ combinators "commute" with reassociation. In other words, it expresses that the type–level associativity of $+$ is properly reflected by the properties of $\oplus$. The last two equivalences show how composition of associativity combinators interact together.

The bottom line in Fig. 7 is actually a linear restatement of the famous "pentagon diagram" stating a particular coherence condition for monoidal categories [49]. To make the relation between Π as a language and the language of category theory, the figure below displays the same morphism but in categorical terms.

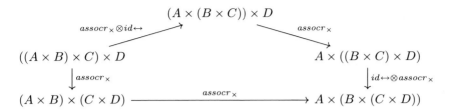

The constructs in Fig. 8 are the basic coherence for *dist, distl, factor,* and *factorl*: the type–level distribution and factoring has to commute with the level–1 $\oplus$ and $\otimes$.

The constructs in Fig. 9 express various properties of composition. The first two says that $id{\leftrightarrow}$ is a left and right identity for sequential composition. The next two say that all programs are reversible, both on the left and the right: running c and then its reverse (! c) is equivalent to the identity, and the same for doing ! c first then c. The last line say that there is an identity level–2 combinator, a sequential composition, and that level–2 equivalence respects level–1 sequential composition $\odot$.

The constructs in Fig. 10 may at first blush look similarly straightforward, but deserve some pause. One obvious question: What is the point of $c_0 : 0 \leftrightarrow 0$, isn't that just the identity combinator $id{\leftrightarrow}$ for $A = 0$ (as defined in Fig. 1)? Operationally, c_0 is indeed indistinguishable from $id{\leftrightarrow}$. However, there are multiple syntactic ways of writing down combinators of type $0 \leftrightarrow 0$, and the first combinator in Fig. 10 applies to all of them uniformly. This is another subtle aspect of coherence: all reasoning must be valid for all possible models, not just the one we have in mind. So even though operational reasoning may suggest that some relations *may* be true between combinators, it can also mislead. The same reasoning applies to $c_1 : 1 \leftrightarrow 1$. The first 8 combinators can then be read as basic coherence for unit introduction and elimination, in both additive and multiplicative cases.

The last two capture another simple idea, related to swapping: eliminating a unit on the left is the same as first swapping then eliminating on the right (both additively and multiplicatively). As a side note, these are not related to *commutativity*, but rather come from one of the simplest coherence condition for braided monoidal categories. In other words, it reflects the idempotence of *swap$_+$* and *swap$_\times$* rather than the commutativity of $\oplus$ and $\otimes$.

The first two equivalences in Fig. 11 reflect the basic coherence between level–0 swapping and the level–1 combinator actions. The next four arise because of interactions between (additive and multiplicative) level–1 associativity and swapping. In other words, they arise as critical pairs. For example, the first expresses that the two ways of going from $(A \oplus B) \oplus C$ to $B \oplus (C \oplus A)$ are equivalent, with the second saying that the reverse (i.e., the results of applying !) also gives equivalent programs. The last two say the same but for the multiplicative structure.

The constructs in Fig. 12 express how unit elimination "in the middle" can be expressed either as operating on the right or, (after reassociation) on the left.

The constructs in Fig. 13 are significantly more subtle, as they deal with combinators involving 0, aka an impossibility. For example,

$$(c \otimes id \leftrightarrow_0) \odot absorbl \Leftrightarrow absorbl \odot id \leftrightarrow_0$$

(where we have explicitly annotated the types of $id \leftrightarrow$ for increased clarity) tells us that of the two ways of transforming from $t_1 * 0$ to 0, namely first doing some arbitrary transformation c from t_1 to t_2 and (in parallel) leaving 0 alone then eliminating 0, or first eliminating 0 then doing the identity (at 0), are equivalent. This is the "naturality" of $absorbl$. One item to note is the fact that this combinator is not irreducible, as the $id \leftrightarrow$ on the right can be eliminated. But that is actually a property visible at an even higher level (which we will not touch in this paper). The next three are similarly expressing the naturality of $absorbr$, $factorzl$, and $factorzr$.

The next combinator, $absorbr \Leftrightarrow absorbl$, is particularly fascinating: while it says something simple—that the two obvious ways of transforming $0 * 0$ into 0, namely absorbing either the left or right 0—it implies something subtle. A straightforward proof of $absorbl$ which proceeds by saying that $0 * t$ cannot be inhabited because the first member of the pair cannot, is not in fact equivalent to $absorbr$ on $0 * 0$. However, if we instead define $absorbl$ to "transport" the putative impossible first member of the pair to its (equally impossible) output, then these do form equivalent pairs. The next few in Fig. 13 also express how $absorbr$ and $absorbl$ interact with other combinators. As seen previously, all of these arise as critical pairs. What is much more subtle here is that the types involved often are asymmetric: they do not have the same occurrences on the left and right. Such cases are particularly troublesome for finding normal forms. Laplaza [46] certainly comments on this, but in mostly terse and technical terms. Blute et al. [42] offer much more intuitive explanations.

The constructs in Figs. 14 and 15 relating associativity and distributivity, and commutativity and distributivity, have more in common with previous

sets of combinators. They do arise from nontrivial critical pairs of different
ways of going between equivalent types. The last one of Fig. 14 is particularly daunting, involving a sequence of three combinators on the left and six
on the right.

4.3 Operational semantics

There are two different interpretations for an operational semantics for the
language of equivalences:

1. Mimicking closely the one in Section 3.3, and thus finding explicit
   homotopies between the functions induced by the operational semantics
   of the level-1 combinators.
2. Treating things more syntactically, and interpreting the combinators as
   program transformations.

A previous paper [32] explores the first interpretation in depth. There one
can find a definition of "equivalences of equivalences," which as the base of
that interpretation.

Here we will focus instead of the syntactic interpretation as program
transformers. This results in a function:

$$eval_1 \; : \; \{t_1 \; t_2 \; : \; U\}\{c_1 \; c_2 \; : \; t_1 \; \leftrightarrow \; t_2\}(ce \; c_1 \Leftrightarrow c_2) \longrightarrow (t_1 \; \leftrightarrow \; t_2) \quad (6)$$

This function is "deeply dependent": given the type of the rewrite ce to
apply, both the input c_1 and output c_2 are almost entirely determined! Let
us take for example the second combinator in Fig. 8:

$$(c_1 \otimes (c_2 \oplus c_3)) \odot distl \Leftrightarrow distl \odot ((c_1 \otimes c_2) \oplus (c_1 \otimes c_3))$$

which we can name distl⇔l. Interpreting this as a rewrite from the program
on the left to the one on the right requires "pattern matching" on the left
structure which contains three arbitrary combinators, from which we can
reconstruct the program on the right. Rewrites such as distl⇔l are one-step
rewrites, in the same way that *distl* is a constant of the base term language
of Π. There is one additional wrinkle. There is naturally an opposite combinator, which interprets the above from right to left; let us call it distl⇔r. It
would appear to require *nonlinear pattern matching* since the right-hand side
contains c_1 twice. That is, however, not the case! The definition of distl⇔r
has five implicit arguments, three of which are c_1, c_2, and c_3, which then
completely force the "shape" of the overall pattern. Thus the mere mention
of distl⇔r is enough to resolve the apparent use of a nonlinear pattern. This is
why *eval_1* was called "deeply dependent" above: once the name of the combinator is given, the rest follows.

If all expressible transformations were single-step only, this would hardly justify calling this an "operational semantics," as we would hardly have a programming language. However, level-2 of Π has combinators as well: two are in Fig. 6 and two are in Fig. 9. The most interesting one is "sequential composition," which is the middle one at the bottom of Fig. 9. Since $\Leftrightarrow$ represents an equivalence, sequential composition in this context is the same as transitivity of equivalences, as thus we have chosen to name this trans$\Leftrightarrow$. When evaluating trans$\Leftrightarrow$, we could cheat: we know that the eventual answer must be, and we could just return that. But this is not operational in any real sense, as that skips over the intermediate steps. We would like to be able to "trace" the rewrite. Thus the evaluation of trans$\Leftrightarrow$ $r_0 r_1$ where $r_0 : c_0 \leftrightarrow c_1$ and $r_1 : c_1 \leftrightarrow c_2$ should apply $eval_1$ to both r_0 and r_1. Furthermore, after applying r_0, we should be able to witness that the result is indeed c_1, so that we may continue. This last requirement forces us to define a new function, mutually recursively with $eval_1$, for this task:

$$exact \ : \ \{t_1 \ t_2 \ : \ U\}\{c_1 \ c_2 \ : \ t_1 \ \leftrightarrow \ t_2\}(ce \ c_1 \Leftrightarrow c_2) \rightarrow eval_1 \ ce \equiv c_2 \quad (7)$$

If we are careful in our construction of $eval_1$, the definition of $exact$ is quite straightforward, i.e., almost all cases are immediately provable by reflexivity.

   This then lets us define the trans$\Leftrightarrow$ $r_0 \ r_1$ case properly: we first evaluate r_0 and get a result combinator, witness that this result type is indeed exactly what we expect, and proceed to evaluate r_1 where we specify that the r_1's left-hand side must be $eval_1 \ r_0$; we can use the Agda keyword rewrite to make this match c_2 "on the nose" (otherwise the call would be ill-typed). This then forces us to use rewrite also in the implementation of the trans$\Leftrightarrow$ case in $exact$.

   The other three combinators are much simpler, as simple recursive calls are sufficient.

4.4 Example

We can now illustrate how this all works with a small example. Consider a circuit that takes an input type consisting of three values $\overset{\frown}{a \ \ b \ \ c}$ and swaps the leftmost value with the rightmost value to produce $\overset{\frown}{c \ \ b \ \ a}$. We can implement two such circuits using our Agda library for Π:

```
swap-fl1 swap-fl2 : {a b c : U} → PLUS a (PLUS b c) ↔ PLUS c (PLUS b a)
swap-fl1 = assocl₊ ⊙ swap₊ ⊙ (id↔ ⊕ swap₊)

swap-fl2 = (id↔ ⊕ swap₊) ⊙
           assocl₊ ⊙
           (swap₊ ⊕ id↔) ⊙
           assocr₊ ⊙
           (id↔ ⊕ swap₊)
```

The first implementation rewrites the incoming values as follows:

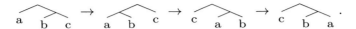

The second implementation rewrites the incoming values as follows:

The two circuits are extensionally equal. Using the level-2 isomorphisms we can *explicitly* construct a sequence of rewriting steps that transforms the second circuit to the first.

We write such proofs in an equational style: in the left column, we have the current combinator which is equivalent to the first one, and in the right column, the justification for that equivalence. The joining combinator is syntactic sugar for trans⇔. The transformation could be written (using trans⇔) by just giving all the pieces in the right-hand column—but such transformations are very hard for humans to understand and follow.

The proof can be read as follows: the first three lines "refocus" from a right-associated isomorphism onto the (left-associated) composition of the first 3 isomorphisms; then apply a complex rewrite on these (the "hexagon" coherence condition of symmetric braided monoidal categories); this exposes two inverse combinators next to each other—so we have to refocus on these to eliminate them; we finally reassociate to get the result.

```
swap-fl2⇔swap-fl1 : {a b c : U} → swap-fl2 {a} {b} {c} ⇔ swap-fl1
swap-fl2⇔swap-fl1 =
  ((id↔ ⊕ swap₊) ⊙ assocl₊ ⊙ (swap₊ ⊕ id↔) ⊙ assocr₊ ⊙ (id↔ ⊕ swap₊))       ⇔⟨ id⇔ □ assoc⊙l ⟩
  ((id↔ ⊕ swap₊) ⊙ (assocl₊ ⊙ (swap₊ ⊕ id↔)) ⊙ assocr₊ ⊙ (id↔ ⊕ swap₊))     ⇔⟨ assoc⊙l ⟩
  (((id↔ ⊕ swap₊) ⊙ assocl₊ ⊙ (swap₊ ⊕ id↔)) ⊙ assocr₊ ⊙ (id↔ ⊕ swap₊))     ⇔⟨ assoc⊙l □ id⇔ ⟩
  ((((id↔ ⊕ swap₊) ⊙ assocl₊) ⊙ (swap₊ ⊕ id↔)) ⊙ assocr₊ ⊙ (id↔ ⊕ swap₊))   ⇔⟨ hexagonl⊕r □ id⇔ ⟩
  (((assocl₊ ⊙ swap₊) ⊙ assocl₊) ⊙ assocr₊ ⊙ (id↔ ⊕ swap₊))                  ⇔⟨ assoc⊙r ⟩
  ((assocl₊ ⊙ swap₊) ⊙ assocl₊ ⊙ assocr₊ ⊙ (id↔ ⊕ swap₊))                    ⇔⟨ id⇔ □ assoc⊙l ⟩
  ((assocl₊ ⊙ swap₊) ⊙ (assocl₊ ⊙ assocr₊) ⊙ (id↔ ⊕ swap₊))                  ⇔⟨ id⇔ □ (linv⊙l □ id⇔) ) ⟩
  ((assocl₊ ⊙ swap₊) ⊙ id↔ ⊙ (id↔ ⊕ swap₊))                                  ⇔⟨ id⇔ □ idl⊙l ⟩
  ((assocl₊ ⊙ swap₊) ⊙ (id↔ ⊕ swap₊))                                        ⇔⟨ assoc⊙r ⟩
  ((assocl₊ ⊙ swap₊) ⊙ (id↔ ⊕ swap₊)) □)
```

4.5 Internal language

Recalling that the λ-calculus arises as the internal language of Cartesian Closed Categories (Elliott [51] gives a particularly readable account of this), we can think of Π in similar terms, but for symmetric Rig Groupoids instead. For example, we can ask what does the derivation in Section 4.4 represent? It is actually a "linear" representation of a 2-categorial commutative diagram! In fact, it is a painfully verbose version thereof, as it includes many *refocusing* steps because our language does not build associativity into its syntax. Categorical diagrams usually do. Thus if we rewrite the example in diagrammatic form, eliding all uses of associativity, but keeping explicit uses of identity transformations, we get that swap-fl2⇔swap-fl1 represents

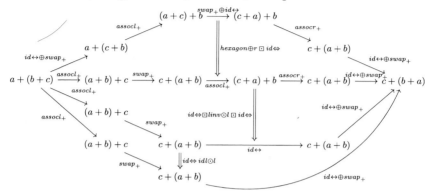

For some, the above diagram will be clearer—it is only three layers high rather than nine! Others will prefer the more programmatic feel of the original definition.

We would be remiss in letting the reader believe that the above is "the" categorical diagram that would be found in categorical textbooks. Rather, congruence would be used to elide the $id⇔$. Furthermore, the various arrows would also be named differently—our $assocl_+$ is often named α, $assocr_+$ is α^{-1}, $swap_+$ is B (always with subscripts). And the two steps needed to remove inverses (i.e., first canceling inverse arrows, then removing the resulting identity arrow "in context") are often combined into one. Here we'll simply name this operation *cancel*, which could be programmed as a defined function over Π level-2. The result would then be the much simpler

In other words, each (nonrefocusing) line of the proof of swap-fl2⇔swap-fl1 is a complete path from left to right in each diagram above, and the annotation on the right-hand side becomes the natural transformation (denoted by vertical ⇒) justifying the move to the next line. The first diagram uses lines 1, 4, 7, 8 in full; the second diagram collapses 7 and 8 into one, as well as not duplicating parts which are related by $id ↔$.

5. Further thoughts and conclusions

We conclude with a collection of open problems and avenues for further research.

5.1 Richer data: Infinite sets and topological spaces

The three languages we discussed only deal with the finite spaces built from 0, 1, sums, and products. Programming practice, logic, and mathematics all deal with richer spaces including inductive types (e.g., the natural numbers, sequences, and trees), functions, and graphs. Extending Π to such domains is possible but only after one refines the notions of reversibility and conservation of information. One approach is to use *partial isomorphisms* that may be undefined on such inputs [30, 52]. Another more speculative approach is to build such spaces, topologically, based on novel type constructions such as negative, fractional, or even imaginary types [53, 54].

5.2 Information effects

A computational model that enforces the principle of conservation of information is arguably *richer* than a conventional model that cannot even express the notion of information. Practically the conventional model is easily recovered by simply adding constructs that intentionally and explicitly create or erase information. Such constructs allow one to recover the classical perspective with the added advantage that it is possible to reason about such creation and erasure of information using type and effect systems, monads, or arrows [52, 55].

An interesting application of such an idea is in the field of *information-flow security*. To make this idea concrete, consider a tiny 2–bit password = "10" and the associated password checker:

```
check-password (guess) =
  guess == "10"
```

One can ask how much information is leaked by this program assuming the attacker has no prior knowledge except that the password is 2 bits, i.e., the four possible 2-bits are equally likely. If the attacker guesses "10" (with probability 1/4) the password (2 bits) is leaked. If the attacker guesses one of the other choices (with probability 3/4) the number of possibilities is reduced from 4 to 3, i.e., the attacker learns $\log 4 - \log 3$ bits of information. So in general the attacker learns:

$$1/4*2 + 3/4(\log 4 - \log 3)$$
$$= 1/4\log 4 + 3/4\log 4/3$$
$$= -1/4\log 1/4 - 3/4\log 3/4$$
$$\sim 0.8 \text{ bits in the first probe}$$

This is a significant amount of information. But of course this is only because the password is so short: if the password was eight restricted ASCII characters (6 bits), the attacker would only learn 0.00001 bits in the first probe.

An alternative formulation of the problem is to view the input as a random variable with four possibilities and a uniform distribution (i.e., with 2 bits of information) and the output as another random variable with four possibilities but with the distribution $\{(\textit{True}, 1/4), (\textit{False}, 3/4)\}$ which contains 0.8 bits of information. Thus 2 input bits of information were given to the password checker and only 0.8 were produced. Where did the 1.2 bits of information go? By the Landauer Principle, these 1.2 bits must be accounted by an *implicit erasure* operation in the program. By writing the password checker in an extension of Π, the erasure construct becomes explicit and the information leak becomes exposed in the syntactic structure of the program [52].

5.3 Theseus and quantum control

The Π family of languages semantically captures the principles of reversibility and conservation of information. As a programming language it has some mixed properties: small programs are relatively easy to write; for some special classes of programs, it is even possible to define a methodology to write large Π programs, including a meta–circular interpreter for Π [56]. In general, however, the point-free style of combinators used in Π becomes awkward and a new approach appears more suitable. To that end, we note that Π encodes the most elementary control structure in a programming language—which is the ability to conditionally execute one of several possible code fragments—using combinators. Expressing such an abstraction using combinators or even predicates and nested **if**-expressions makes it difficult for both humans and compilers to write, understand, and reason about the control flow structure of the program. Instead, in modern functional languages, this control flow paradigm is elegantly expressed using *pattern matching*. This approach yields code that is not only more concise and readable but also enables the compiler to easily verify two crucial properties: (i) nonoverlapping patterns and (ii) exhaustive coverage of a datatype using a collection of patterns. Indeed most compilers for functional languages perform these checks, warning the user when they are violated. At a more fundamental level, e.g., in type theories and proof assistants, these properties are actually necessary for correct reasoning about programs. Our insight is that these properties, perhaps surprisingly, are sufficient to produce a simple and intuitive first-order reversible programming language which we call *Theseus*.

We provide a small illustrative example, written in a Haskell-like syntax. Fig. 16 gives the skeleton of a function f that accepts a value of type `Either Int Int`; the patterns on the left-hand side exhaustively cover every possible incoming value and are nonoverlapping. Similarly, Fig. 17 gives the skeleton for a function g that accepts a value of type `(Bool,Int)`; again the patterns on the left-hand side exhaustively cover every possible incoming value and are nonoverlapping. Now we claim that since the types `Either Int Int` and `(Bool,Int)` are isomorphic, we can combine the patterns of f and g into *symmetric pattern matching clauses* to produce a reversible function between the types `Either Int Int` and `(Bool,Int)`. Fig. 18 gives one such function; there, we suggestively use `<->` to indicate that the function can be executed in either direction. This reversible function is obtained by simply combining the nonoverlapping exhaustive patterns on the two sides of a clause. In order to be well-formed in either direction, these clauses are subject to the

```
f :: Either Int Int -> a
f (Left 0)     = undefined
f (Left (n+1)) = undefined
f (Right n)    = undefined
```

Fig. 16 A skeleton.

```
g :: (Bool,Int) -> a
g (False,n)  = undefined
g (True,0)   = undefined
g (True,n+1) = undefined
```

Fig. 17 Another skeleton.

```
h :: Either Int Int <-> (Bool,Int)
h (Left 0)     = (True,0)
h (Left (n+1)) = (False,n)
h (Right n)    = (True,n+1)
```

Fig. 18 An isomorphism.

constraint that each variable occurring on one side must occur exactly once on the other side (and with the same type). Thus it is acceptable to swap the second and third right-hand sides of h but not the first and second ones. With some additional work, it is possible to extend Theseus to a full-fledged reversible programming language [57]. With just one additional insight, Theseus can be extended with superpositions and becomes a quantum programming language [58].

5.4 Quantum speedup

A rather remarkable but somehow overlooked paper is "Quantum speedup and Categorical Distributivity" by Peter Hines [59]. Here he shows that the heart of Shor's algorithm can be reduced to an operation $!^N()$ (expressible in Π), which can be expressed, via a factorization, in an exponentially faster manner. The key to this efficient factorization is exactly the coherence conditions of Laplaza [46], which also feature prominently in our work. Proving his key Lemma 2 in Π could be quite instructive in revealing which level-2 combinators are crucial for this result.

5.5 Summary

The entire edifice of computer science including its mainstream models of computations, programming languages, and logics is founded on *classical physics*. While much of the world phenomena can be approximated with classical physics, we are reaching a revolutionary period of quantum technology that challenges many of the classical assumptions. It remains to be seen how computer science will adapt to this quantum revolution but we believe that additional physical principles inspired by quantum mechanics will have to be embraced in our computational thinking. This paper focused on one such principle—*conservation of information*—and explored some of its exciting implications to the field of computer science.

Acknowledgments

We would like to thank the numerous students and colleagues who participated in various aspects of this research and who provided valuable feedback and constructive criticism. This material is based upon work supported by the National Science Foundation under Grant No. 1936353.

References

[1] T. Toffoli, Reversible computing, in: Proceedings of the Seventh Colloquium on Automata, Languages and Programming, Springer-Verlag, 1980, pp. 632–644.
[2] R. Feynman, Simulating physics with computers, Int. J. Theor. Phys. 21 (1982) 467–488.

[3] H. Leff, R. Rex, Maxwell's Demon: Entropy, Information, Computing, Princeton University Press, Princeton, NJ, 1990.

[4] R. Landauer, Irreversibility and heat generation in the computing process, IBM J. Res. Dev. 5 (3) (1961) 183–191.

[5] A. Bérut, A. Arakelyan, A. Petrosyan, S. Ciliberto, R. Dillenschneider, E. Lutz, Experimental verification of Landauer's principle linking information and thermodynamics, Nature 483 (2012) 187–189.

[6] A. Turing, On computable numbers, with an application to the entscheidungsproblem, in: 2, Proceedings of the London Mathematical Society, vol. 42, 1936, pp. 230–265. Available from: http://www.abelard.org/turpap2/tp2-ie.asp.

[7] A. Church, The Calculi of Lambda-Conversion, Ann. Math. Stud., vol. 6, Princeton University Press, Princeton, 1951 (second printing, first appeared 1941).

[8] C.E. Shannon, A mathematical theory of communication, Bell Syst. Tech. J. 27 (1948) 379–423. 623–656.

[9] C.H. Bennett, Logical reversibility of computation, IBM J. Res. Dev. 17 (6) (1973) 525–532.

[10] C.H. Bennett, R. Landauer, The fundamental physical limits of computation, Sci. Am. 253 (1) (1985) 48–56.

[11] C.H. Bennett, Notes on the history of reversible computation, IBM J. Res. Dev. 32 (1) (2010) 16–23.

[12] C.H. Bennett, Notes on Landauer's principle, reversible computation, and Maxwell's Demon, Stud. Hist. Philos. Sci. Part B Stud. Hist. Philos. Mod. Phys. 34 (3) (2003) 501–510.

[13] H.G. Baker, NREVERSAL of fortune–the thermodynamics of garbage collection, in: Proceedings of the International Workshop on Memory Management, Springer-Verlag, 1992, pp. 507–524.

[14] J. Baez, M. Stay, Physics, topology, logic and computation: a Rosetta Stone, New Struct. Phys. (2011) 95–172.

[15] P. Malacaria, F. Smeraldi, The thermodynamics of confidentiality, in: CSF, 2012, pp. 280–290.

[16] J.D. Bekenstein, Universal upper bound on the entropy-to-energy ratio for bounded systems, Phys. Rev. D 23 (2) (1981) 287–298, https://doi.org/10.1103/PhysRevD.23.287.

[17] J.-Y. Girard, Truth, modality and intersubjectivity, Math. Struct. Comput. Sci. 17 (6) (2007) 1153–1167, https://doi.org/10.1017/S0960129507006342.

[18] H.G. Baker, Lively linear Lisp—look ma, no garbage! SIGPLAN Not. 27 (8) (1992) 89–98.

[19] A. Peres, Reversible logic and quantum computers, Phys. Rev. A 32 (6) (1985).

[20] M.P. Frank, Reversibility for efficient computing (Ph.D. thesis), Massachusetts Institute of Technology, 1999.

[21] A. van Tonder, A lambda calculus for quantum computation, SIAM J. Comput. 33 (5) (2004) 1109–1135.

[22] W.E. Kluge, A reversible SE(M)CD machine, in: International Workshop on Implementation of Functional Languages, Springer-Verlag, 2000, pp. 95–113.

[23] L. Huelsbergen, A logically reversible evaluator for the call-by-name lambda calculus, Int. J. Complex Syst. 46 (1996).

[24] V. Danos, J. Krivine, Reversible communicating systems, Concurrency Theory. Lecture Notes in Computer Science vol. 3170 (2004) 292–307.

[25] T. Yokoyama, R. Glück, A reversible programming language and its invertible self-interpreter, in: PEPM, ACM, 2007, pp. 144–153.

[26] T. Yokoyama, H.B. Axelsen, R. Glück, Principles of a reversible programming language, in: Conference on Computing Frontiers, ACM, 2008, pp. 43–54.

[27] S.-C. Mu, Z. Hu, M. Takeichi, An injective language for reversible computation, in: MPC, 2004, pp. 289–313.

[28] S. Abramsky, A structural approach to reversible computation, Theor. Comput. Sci. 347 (2005) 441–464.

[29] A. Di Pierro, C. Hankin, H. Wiklicky, Reversible combinatory logic, MSCS 16 (4) (2006) 621–637.

[30] W.J. Bowman, R.P. James, A. Sabry, Dagger traced symmetric Monoidal categories and reversible programming, in: Workshop on Reversible Computation, 2011.

[31] R.P. James, A. Sabry, Information effects, in: Proceedings of the ACM SIGPLAN-SIGACT Symposium on Principles of Programming Languages, ACM, 2012, pp. 73–84.

[32] J. Carette, A. Sabry, Computing with semirings and weak rig groupoids, in: ESOP 2016, Springer, Berlin, Heidelberg, 2016, pp. 123–148.

[33] A.J.G. Hey, Feynman and Computation: Exploring the Limits of Computers, Perseus Books, Cambridge, MA, USA, 1999, ISBN: 0-7382-0057-3.

[34] E. Fredkin, T. Toffoli, Conservative logic, Int. J. Theor. Phys. 21 (3) (1982) 219–253.

[35] M. Hasegawa, Recursion from cyclic sharing: traced monoidal categories and models of cyclic lambda calculi, in: TLCA, Springer-Verlag, 1997.

[36] B. Desoete, A. De Vos, Feynman's reversible logic gates, implemented in silicon, in: Proceedings of the Sixth Advanced Training Course on Mixed Design of VLSI Circuits, Technical University Lodz, Napieralski A. (ed.), Krakow, June 1999, 1999, pp. 497–502.

[37] Y. Van Rentergem, A. De Vos, Optimal design of a reversible full adder, IJUC 1 (2005) 339–355.

[38] Univalent Foundations Program, Homotopy Type Theory: Univalent Foundations of Mathematics, Institute for Advanced Study, 2013. http://homotopytypetheory.org/book.

[39] M. Fiore, Isomorphisms of generic recursive polynomial types, in: Proceedings of the ACM SIGPLAN-SIGACT Symposium on Principles of Programming Languages, ACM, 2004, pp. 77–88.

[40] M.P. Fiore, R. Di Cosmo, V. Balat, Remarks on isomorphisms in typed calculi with empty and sum types, Ann. Pure Appl. Logic 141 (1–2) (2006) 35–50.

[41] I. Mackie, The geometry of interaction machine, in: Proceedings of the ACM SIGPLAN-SIGACT Symposium on Principles of Programming Languages, 1995, pp. 198–208.

[42] R.F. Blute, J.R.B. Cockett, R.A.G. Seely, T.H. Trimble, Natural deduction and coherence for weakly distributive categories, J. Pure Appl. Algebra 113 (3) (1996) 229–296, https://doi.org/10.1016/0022-4049(95)00159-X.

[43] P. Selinger, A survey of graphical languages for Monoidal categories, in: B. Coecke (Ed.), New Structures for Physics, Lecture Notes in Physics, vol. 813, Springer Berlin, Heidelberg, 2011, pp. 289–355.

[44] J. Carette, C.-H. Chen, V. Choudhury, A. Sabry, From reversible programs to univalent universes and back, Electron. Notes Theor. Comput. Sci. 336 (2018) 5–25, https://doi.org/10.1016/j.entcs.2018.03.013. The 33rd Conference on the Mathematical Foundations of Programming Semantics (MFPS XXXIII).

[45] M. Hofmann, T. Streicher, The groupoid interpretation of type theory, in: Venice Festschrift, 1996, pp. 83–111.

[46] M.L. Laplaza, Coherence for distributivity, in: G.M. Kelly, M. Laplaza, G. Lewis, S. Mac Lane (Eds.), Coherence in Categories, Lecture Notes in Mathematics, vol. 281, Springer Verlag, Berlin, 1972, pp. 29–65, ISBN: 978-3-540-05963-9, https://doi.org/10.1007/BFb0059555.

[47] G.M. Kelly, Coherence theorems for lax algebras and for distributive laws, in: G.M. Kelly (Ed.), Category Seminar, Springer, Berlin, Heidelberg, 1974, pp. 281–375, ISBN: 978-3-540-37270-7.

[48] S. Mac Lane, Topology and logic as a source of algebra, Bull. Am. Math. Soc. 82 (1976) 1–40.

[49] G.M. Kelly, S. Mac Lane, Coherence in closed categories, J. Pure Appl. Algebra 1 (1) (1971) 97–140, https://doi.org/10.1016/0022-4049(71)90013-2.

[50] T.M. Fiore, P. Hu, I. Kriz, Laplaza sets, or how to select coherence diagrams for pseudo algebras, Adv. Math. 218 (6) (2008) 1705–1722, https://doi.org/10.1016/j.aim.2007.05.001.

[51] C. Elliott, Compiling to categories, Proc. ACM Program. Lang. 1 (ICFP) (2017), https://doi.org/10.1145/3110271. http://conal.net/papers/compiling-to-categories.

[52] R.P. James, A. Sabry, Information effects, in: Proceedings of the ACM SIGPLAN-SIGACT Symposium on Principles of Programming Languages, 2012.

[53] A. Blass, Seven trees in one, J. Pure Appl. Algebra 103 (1–21) (1995).

[54] R.P. James, The computational content of isomorphisms (Ph.D. thesis), Indiana University, 2013.

[55] C. Heunen, R. Kaarsgaard, M. Karvonen, Reversible effects as inverse arrows, in: MFPS, 2018.

[56] R.P. James, A. Sabry, Isomorphic interpreters from small-step abstract machines, in: Reversible Computation, 2012.

[57] R.P. James, A. Sabry, Theseus: a high-level language for reversible computation, in: Reversible Computation, 2014. Booklet of work-in-progress and short reports.

[58] A. Sabry, B. Valiron, J.K. Vizzotto, From symmetric pattern-matching to quantum control, in: C. Baier, U. Dal Lago (Eds.), Foundations of Software Science and Computation Structures, Springer International Publishing, Cham, 2018, pp. 348–364, ISBN: 978-3-319-89366-2.

[59] P. Hines, Quantum speedup and categorical distributivity, in: Computation, Logic, Games, and Quantum Foundations. The Many Facets of Samson Abramsky, Springer, 2013, pp. 122–138.

About the authors

Jacques Carette received his PhD in Mathematics from the Universite de Paris-Sud, Orsay, France in 1997. His research interests include generative and meta programming, new forms of programming, and mechanized mathematics.

Roshan P. James completed his PhD in Computer Science from Indiana University and has worked at Jane Street Capital and Google.

Amr Sabry received his PhD in Computer Science from Rice University in 1994. His research interests are in the semantics, logical foundations, and implementations of programming languages.

CHAPTER THREE

WSNs in environmental monitoring: Data acquisition and dissemination aspects

Zhilbert Tafa
University of Montenegro, Podgorica, Montenegro

Contents

Abstract

The growth in world population, industrialization, and urbanization, are inducing harmful effects on living environment. Current air and water pollution trends indicate the emergent need for actions toward systematic and continuous identification and management of the pollution sources. Conventional environmental monitoring systems are based on manual periodic on-site sampling, typically by using expensive sparsely deployed instruments. As such, these systems cannot appropriately follow the pollutants' spatiotemporal distribution. On the other hand, the advances in low-cost sensors, embedded systems and low-power wireless technologies, qualify the Internet of Things (IoT) information systems to be the natural technological response to the problem solution. With Wireless Sensor Networks (WSNs)—based infrastructure for data acquisition and dissemination, these systems have the potential to perform (near) real time and continuous pollution measurements, even in hardly accessible and harsh environments. This can have a great impact

Advances in Computers, Volume 126
ISSN 0065-2458
https://doi.org/10.1016/bs.adcom.2021.11.010

65

to the on-time detection and/or prediction of the pollution sources and hotspots. However, these systems are still not widely deployed because of the actual technological constraints (such as reliability, accuracy, robustness, etc.). This article presents, analyzes and classifies the current technological efforts toward (near) real time low-cost continuous water and air quality monitoring, while being focused mainly on sensing, processing and communication techniques and algorithms for data acquisition and dissemination. It aims to investigate the need for new technologies in this area, their hardware and communication structure, as well as their potential to replace the existing technologies while facing many challenges. Future research, academic and engineering directions on improving of the actual systems are also proposed.

Abbreviations

6LoWPAN	Ipv6 over low power wireless personal area networks
A/D	analog to digital
ANN	artificial neural networks
AQI	air quality index
AQM	air quality monitoring
CDMA	code division multiple access
CoAP	constrained adaptation protocol
CPU	central processing unit
CSS	community sensor networks
DO	dissolved oxygen
EPA	environmental protection agency
GaAs	gallium arsenide
GPRS	general packet radio service
GPS	global positioning system
GSM	global system for mobile communications
HVAC	heating ventilation and air conditioning
IoT	internet of things
IoUT	internet of underwater things
ISM	industrial scientific medical
LCS	low-cost sensors
LPWAN	low power wide area network
LTE	long term evolution
MAC	media access protocol
MOS	metal oxide semiconductor
NB-IoT	narrowband internet of things
NTU	nephelometric turbidity units
OPC	optical particle counters
ORP	oxygen reduction potential
pH	potential of hydrogen
PM	particulate matter
RAM	random access memory
RF	radio frequency
SAW	surface acoustic wave
SMAC	sensor media access control

SSN	stationary sensor networks
SVM	support vector machine
TCP	transmission control protocol
TDMA	time division multiple access
TDOA	timed difference of arrival
TDS	total dissolved solids
TEOM	tapered element oscillating micro-balance
UDP	user datagram protocol
VOC	volatile organic compounds
VSN	vehicular sensor network
WDS	water distribution system
WHO	World Health Organization
WLAN	wireless local area networks
WQI	water quality index
WQM	water quality monitoring
WSN	wireless sensor network

1. Introduction

Current advances in technology and economy are having a significant impact over the environment, and have led to serious concerns regarding pollution and climate change [1]. For the time being, around 20% of population lacks clean drinking water; the air pollution measured in many cities reaches unhealthy or even hazardous indexes and the soil quality/contamination is greatly influencing the agriculture and its products. Consequently, the continuous environmental monitoring is extremely important to ensure a safe and wealthy life of both humans and artifacts [2].

Conceptually, the environmental monitoring involves quality measures applied to: water, air, and soil. Traditional environmental monitoring systems have been based on periodical manual measurements at predetermined, but not necessarily optimal points of measurements. These relatively rare (e.g., once a month), isolated samplings, which have been typically performed at a few points of interest in a given area, are not sufficient to follow the dynamics of the today's environment pollution over time and space. The measurements based on traditional manual lab-based approach show an inability to conduct trend analysis based on historic data, as data may not be sampled frequently enough for some analyses, and additionally data can be lost at any given time due to the manual processes involved in data collection and recording [3]. Therefore, the development of high-sensitivity multi-parameter, spatially distributed, real-time monitoring systems appear to be of crucial importance in appropriate pollution prevention and/or management. These systems receive a growing interest as environmental technology becomes a key field of sustainable growth worldwide [4].

Typically, WSNs have been involved to the problem solution at the data acquisition and dissemination layer. Wireless Sensor Networks enable low cost and high coverage sensing by using small (usually power autonomous) nodes to sense and wirelessly transmit physical phenomena. As such, they represent the most promising technological solution to real time remote environmental monitoring. The final goal is the construction of the IoT-based information systems which can provide relatively inexpensive, coordinated, real time, high coverage, and intelligent solutions to better manage and protect natural resources. However, depending on the application area, different system architectures are needed to measure different pollution indicators in different environments. For example, water quality measurements and data transmission would be performed differently when measuring water quality in lakes as compared to those performed in water distribution systems-pipes. Also, different sensing and communication requirements are set when the air pollution is measured in the enclosed environments as compared to those when the urban pollution is measured. These systems would often differ both in number and types of the measured parameters as well as in the technology used to acquire and transmit data. In summary, there are no universally accepted approaches or technologies that would match all the requirements. Therefore, various solutions have been designed for various applications.

The aim of this article is to present, analyze and classify the actual low-cost data acquisition and dissemination water and air pollution monitoring systems. It reviews some recent examples from literature and practice, with the aim to highlight the actual strengths and drawbacks of the existing solutions and technologies. By relying on theory and the observations obtained from practically implemented systems, it also aims to identify the key challenges as well as to make some recommendations on the future research trends, application design and possible solutions. In short, the article investigates the needs for new technologies in this area, analyzes their structure (in the framework of data acquisition and dissemination technologies), proposes the optimizations based on the state-of-the-art and theory in this field, and estimates the overall potential of replacing the existing technologies with the low-cost IoT-based applications. Finally, the need for a complete integrated system that would enable the precise and reliable environment data acquisition, mobile remote data visualization, as well as the pollution hotspot identification and prediction, is recognized as one of the future technological goals.

The reminder of this article is structured as follows. The following section gives the overview on the WSN systems and technologies for remote environmental monitoring, mainly focused on their hardware architectures, data transmission aspects, and power consumption/management. In Section 3 the taxonomy, the classification, and basic requirements of the low-cost water quality information systems are presented first. In the proceedings of this section, sensing, processing, and communication aspects of these systems are presented. The last part of this section covers some water quality monitoring (WQM) examples from literature and practice, highlights their strengths and drawbacks, and classifies them on various criteria. Section 4 is organized similarly as Section 3, with the same organizational structure, but with the difference in focus, i.e., it covers air quality monitoring (AQM). Finally, based on the analysis from previous sections, Section 5 discusses the various technological and implementation aspects, from challenges and optimization possibilities to future opportunities.

2. WSNs and their position in environmental monitoring

The concepts of IoT and WSN were introduced a few decades ago but they were still not ready for the implementation due to the technological constraints. Today, small power-autonomous devices have satisfying processing, memory, and transmission capacity to acquire the information from the environment, to provide it to the user, and/or to receive and execute the users' requirements. These devices, which can be of physical dimensions of orders of millimeters squared, can participate in (self-organizing) wireless networks that can sense and transmit data to the remote locations.

The IoT, as a broader concept, involves various technologies for data acquisition and transmission, ranging from the fixed power supplied high resolution sensors and systems to the battery supplied wireless micro-devices. It aims to establish the connection between "things" and Internet, while a "thing" can be a landmark, a monument, clothing, food, vehicle, a refrigerator etc. Typically, small devices-nodes are attached to the "things" to enable the communication between them or to communicate the information to the users (over the data network). In order to connect things from different application areas such as smart homes, smart cities, e-health, military surveillance applications, environmental monitoring, etc. (Fig. 1), IoT systems may include various networking technologies.

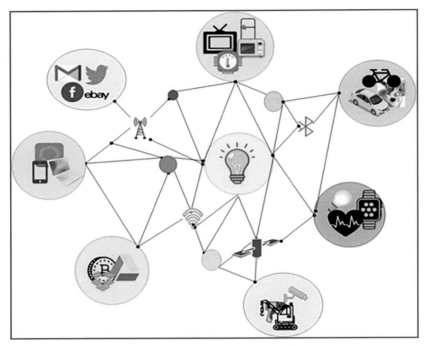

Fig. 1 Examples of IoT elements and applications.

Data communications are converging toward wireless technologies. At the same time, the development of low power embedded systems enables the node's power-autonomous lifetime of order of months or even years. This is why wireless sensor systems have become the technology of choice within the framework of the IoT data acquisition and dissemination infrastructure. A WSN is a network of small power-autonomous devices—nodes which can sense the information from the environment, locally process it, and transmit it wirelessly to the user and/or datacenter. Due to the small physical dimensions, power autonomy, ad hoc installation, and the small overall network cost, WSNs have been used in many IoT-based applications for environmental monitoring.

A typical WSN-based information system for this purpose involves the following concepts:
(a) Data acquisition—sensing, signal conditioning, and data processing.
(b) Data communication—communication technologies, protocols, algorithms, and topologies.
(c) Energy conservation and management approaches.
(d) Data management, analysis, and utilization.

This article is focused on first three issues, although, in fewer details, other system-performance parameters related to these issues (such as: nodes' size, weight and price, programming requirements, design flexibility, nodes' electromechanical robustness, data utilization etc.) will be considered as well. Following this conception, the proceeding of this section will be mainly focused on three aspects: (a) nodes' architecture—hardware composition, (b) network organization and communication aspects, and (c) energy conservation and management. Data visualization and utilization will be shortly considered in the wider context of IoT information systems and during the analyses of the presented examples.

2.1 Node's components, architectures and properties

One of the first steps in the design of WSN-based systems includes selection of the sampling sites, i.e., the nodes' spatial resolution, position, and distribution. Depending on the nature of applications, sampling and data reporting frequency should also be chosen to assure the appropriate temporal resolution. Finally, in accordance to the required spatial and temporal resolution, network coverage, and computational timings, the composition of the smart sensors have to be optimally designed regarding the processing power, energy consumption, physical size, cost, transmission range, etc.

A typical sensor node contains sensor module (with integrated sensing devices and signal conditioning circuits), microcontroller module, data transceiver module, and power supply/management system (Fig. 2). Since many computational, storage, and data transfer capabilities are typically added to the sensing process, sensor nodes are sometimes referred as smart sensors.

A sensor enables for the transformation of the physical event or process (light, concentration of some chemical element, vibrations etc.) into the electrical signal. Usually, these signals are analog. Hence digitalization is needed in order for the outputs to be further processed by a microcontroller. As these functionalities are typically combined in a single module, the sensor array along with the appropriate signal conditioning circuits and A/D converter are usually referred to as a sensing module.

The most important performance indicators of a sensor are its resolution, sensitivity, selectivity, response time, precision, and accuracy. In order for the high sensing/communication coverage and/or redundancy to be achieved, sometimes a greater number of sensors have to be deployed. Additionally, for the system to remain economically feasible, sensors should be of low cost. The "low-cost" attribute refers to the price of the device

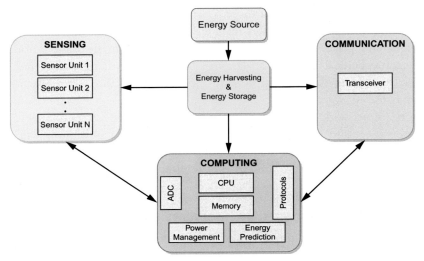

Fig. 2 Main components of a typical smart sensor.

compared to its equivalent professional reference device which is of an order of magnitude more expensive. Sensor resolution is related to the A/D converters and the sensor's output voltage range. It is measured as the smallest step (in Volts) that can be produced by a change in the measured signal. For example, if the voltage range is between $-10\,\text{V}$ and $+10\,\text{V}$, and the A/D converter is of 10 bits, then the sensor's resolution is $20\,\text{V}/2^{10} = 19.5\,\text{mV}$ per A/D count. Sensor's sensitivity presents the smallest amount of value that can be detected by a sensor—i.e., the smallest value at which the sensor reacts with some output. Selectivity, on the other hand describes the ability of a (multi-parameter) sensor to differentiate one measuring parameter from another. For example, a gas sensor is supposed to measure many gas parameters. In this case, the selectivity would describe the sensor's ability to successfully differentiate one gas from another. Response time of a sensor describes the time elapsed between the moment when a specific sensing parameter reaches a specific value, and the moment when a sensor reports that value. The sensor precision is also called repeatability. It describes the ability of the sensor to produce approximately the same output for the same input value of the measured parameter. Finally, the accuracy presents the degree at which the sensor measurements deviate from the predefined absolute established standards in a given application area.

In practice, the sensors' performance parameters are influenced by other parameters as well, such as: drift over time, interference from climate and environmental conditions, mechanical influences, etc. For example, most sensors, especially their zero reference, impose some drift and non-linearity in their readings with regards to the operating temperature. Other conditions and/or biochemical influences such as humidity, biofouling, etc., will also change the sensor's performances. Consequently, sensors should be calibrated and periodically rechecked for precision and accuracy with reference to the available standards. Calibration consists of setting a mathematical model describing the relationship between LCS data and reference measurements [5]. The calibration process involves either establishing the reference point for each sensor from expensive high quality sensors or by using the estimated reference. Both approaches have drawbacks in the system's price and accuracy.

In order to adapt the sensor outputs to the microcontroller circuit, additional electronic components are often needed. This functional block is known as the signal conditioning circuit. And while many devices use the same or similar sensors, manufacturers use a variety of algorithms to convert the sensor output to a concentration of each pollutant [6]. The way the signal is conditioned can significantly influence the sensor's overall precision and accuracy. Finally, the sensors' output data should be compatible with the data representation standards that are widely used in computer systems and Internet.

The selection of the appropriate processing unit is the next step in an application design. The processing unit is responsible for the management of hardware component functions and data processing. It typically contains a microcontroller, Random Access Memory (RAM) for temporary data storage, long term flash memories for program and data storage etc. The microcontroller enables in-node processing, aggregation, filtering, decision making and algorithm implementation (if any). A node can process data and measure specific values at given points, or it can collaborate with other nodes by exchanging/sharing data in order to characterize a specific region with regard to a specific value, e.g., in this manner a WSN can find an average value of the air pollution for a given region. It may also communicate with the user—it may receive the requests from the users and perform actions. For example, a user can query for the energy level of the nodes or may request for the change of the sampling rate in order to improve the measurement quality or to reduce the power consumption. However, environmental monitoring is generally a process of low complexity; therefore, the advanced microcontrollers are not required and are mostly avoided because of their higher power consumption and price.

After signal processing, the data are forwarded to the transceiver for wireless transmission. And while sensing and processing are simpler tasks, network organization and data dissemination remain the most challenging parts of the system design, mainly because of the following issues. First, due to the nodes' mobility, weather conditions, etc., the network may operate at the conditions of unstable links and limited output power of the transceivers. Even in these circumstances, the required sensing coverage and connectivity should be maintained. Second, to achieve the required coverage with as less energy consumption as possible, nodes typically need to establish and maintain short-range multi-hop communications. In such topologies, where data from various sensors are typically aggregated and routed through specific nodes, energy holes can appear. This phenomenon happens typically at the cluster heads or at a given backbone node. Precisely, due to the high data load and long operation in active mode, the energy of the cluster heads may get quickly drained, when they would eventually fail to operate. Also, the multi-hop data routing protocols for densely deployed WSNs with unstable links and possibly great collision, are much more complex than ordinary routing protocols (of the wire line networks). Finally, the nodes should conserve self-healing and self-organization capability even if the deployment is quasi random. An example of quasi randomly deployed and highly dynamic WSN-based network is Vehicular Senor Network (VSN).

Having in mind the aforementioned limitations, traditional low-cost IoT-based solutions for long term in-situ remote monitoring rely typically on: (a) short range wireless communications for transferring data to the nearby gateway device, and on (b) long range (wireless or wired) communication technologies for further data transmission to the data center. However, as it will be shown in the proceeding of this article, other network architectures that have been proposed in literature can also perform well in specific situations. However, in most of the cases so far, both of the following types of wireless communications would be used: short range communications (ZigBee, Bluetooth, WiFi, etc.) and Wireless Wide Area Network (WWAN) links (GPRS, 34/4G etc.).

Differently from the aforementioned wireless technologies, recent developments in the field of data communications for IoT have brought solutions that compete with the traditional WSN-based paradigm. In the last 10 years, new Low Power Wide Area Networks (LPWAN) have emerged. The LPWAN solutions are based on the narrowband communications between sensor nodes and base stations. Main advantage of these technologies over the existing ones is that they enable long range communication (of order of tens of km) with long battery life (of order of years). The leading LPWAN technologies that have survived so far are LoRa, Sigfox, and Narrow Band IoT (NB-IoT).

Both traditional and LPWAN technologies for IoT-based solutions will be described in more details in Section 2.2, while a number of implementations will be incorporated into the respective subsections on water and air quality example systems.

Finally, node's architecture (presented in Fig. 2) depends on the choice of the hardware components, their compatibility and easiness of integration/ programming, the hardware integration manner, design flexibility, etc. The components of WSN nodes can be integrated differently, depending on the application requirements and cost. A classification with regards to the node's composition is given in Table 1.

The nodes that belong to the class H.1 are typically more robust and resistant to the environmental and installation conditions such as: moisture, high or low temperature, electromechanical influences, etc. These nodes are manufactured and tested to work at specific environment conditions. They are usually pre-calibrated against a reference standard and can be remotely rechecked. They are also easier to program and to implement, with some of them being (almost) plug-and-play, such as Libelium Waspmote Plug and Sense [7], MicaZ motes [8], etc. From the user point of view, H.1 systems are out-of-box ready-to-use, while H.2 and H.3 are manually constructed and programmed from the scratch from various components. However, H.1 class nodes are usually expensive and provide relatively small design flexibility as compared to the H.2 and H.3 class nodes. The nodes that belong to the classes H.2 and H.3 can be designed for various purposes,

Table 1 Classification of the node's hardware composition.

Hardware composition	Description
H.1	A standalone device, with integrated sensing, signal conditioning, signal processing, and communication module
H.2	A device of two interconnected boards: (a) the probes with signal conditioning (and A/D converter), (b) signal processing and communication device
H.3	A set of off-the-shelf interconnected devices that separately do sensing, signal conditioning, A/D conversion, data processing (possibly with algorithm implementation), and wireless transmission

i.e., they are not limited to the specific application and/or environment. The designer can easily select and replace the components in accordance to the application requirements. Also, the cost of the whole system composed of H.2 and H.3 nodes is much lower as compared to the H.1 systems. However, when building a system from the scratch, the programming, as well as the implementation and maintenance, become more demanding.

All of the aforementioned classes can be enhanced by alternative power supply sources and energy-harvesting systems.

2.2 Data dissemination

In a typical WSN, a node sends data wirelessly via other (intermediate) nodes, or communicates directly with the data center. Various networking protocols and technologies have been proposed, but there is still no technology that can perfectly map the applications' Quality of Service (QoS) requirements into its technical constraints space. The most critical issue is balancing between the network deployment/coverage, energy efficiency, and the network performances. For example, if a routing protocol is more "energy efficient," due to its low duty cycle or due to its intent to route through the most energy efficient routes (instead through the shortest paths), it might impose the long end to end delays or it can fail to find a specific route. Also, in a network with high sensing coverage as a primary goal, a densely deployed WSN would be required. High wireless network density, on the other hand, increases the probability of collision occurrence. This increases the packet loss probability. To compensate for the packet loss, nodes will usually retransmit. Finally, the retransmissions will cause the unwanted consequences, such as: traffic overhead, increased end-to-end latency, and reduced nodes' lifetime (due to the higher activity period). In order to address these issues, a number of relatively complicated Media Access Control (MAC) protocols have been developed. On the other hand, the more complex is the MAC protocol, the greater the processing power consumption will be.

2.2.1 Traditional wireless technologies for environmental monitoring

Although many different wireless networking technologies and protocols have been proposed, a few of them have received greater attention in specific fields of applications. For example, within the framework of traditional approaches to local data dissemination, Bluetooth 4.0 is mostly oriented toward medical applications while the IEEE 802.15.4e toward the industrial applications [9]. Few successful efforts have also been made on using WLAN (Wireless Local Area Network) modules and BLE (Bluetooth Low Energy) technologies for local data transmission. For example, a study on temperature and humidity monitoring presented in Ref. [10] shows that if WiFi communication is used along with the UDP (User Datagram Protocol) and small enough active/sleep nodes' duty cycling, the network can provide battery lifetime of up to 2 years. In a such implementation, WLAN and BTE technologies would generally outperform the IEEE 802.15.4 network because, while enabling for long enough battery life, they have the additional performance-related advantages over IEEE 802.15.4, such as link stability, small delay and jitter. Besides, WiFi and BTE are already present in most of the mobile IoT devices today (such as smart phones, tablets, etc.). This makes them suitable in the sense of interoperability and integration.

In general, the decision on network architecture and technology is mainly dictated by the required coverage/range and nodes' lifetime, although sometimes other QoS parameters (such as throughput, delay, etc.) might also influence the decision. Typically, if distances from sensing points to the data center or to the gateway device are not very long (shorter than a km), the designers aim to utilize low power (multi-hop) wireless technologies. Otherwise, at a price of the higher power consumption per node, some of the long-range wireless communications would be used. Long range wireless links, such as Long Term Evolution (LTE), 3G etc., have been typically used in combination with the short range technologies—to transmit data from the gateway device to the remote locations. These technologies are also suitable in applications with a single sensing node and a point-to-point connection to the data center, or when the nodes are supplied from the power grid (e.g., nodes for air pollution monitoring mounted on the streetlights).

Although other aforementioned technologies (such as Bluetooth, WiFi, BTE, etc.) have been also used for local data dissemination, most of the applications use ZigBee standard for this purpose. ZigBee protocol stack relies on the IEEE 802.15.4 at the data link and physical layer. It is short range low power technology that operates at 2.4 GHz, 900 MHz, and 868 MHz Industrial Scientific Medical (ISM) bands with data rates varying from 240 kbs to 20 kbps. A Zigbee compliant node could be configured as an end device/sensor node, as a router, or as a coordinator. The end node sends data (via router nodes or directly) toward the sink node. The sink node can be connected directly to the data storage and monitoring station, or/and it can send data through the Internet infrastructure (or directly to the data center) by using some the WWAN technologies. The coordinator node organizes the network topology. The technology supports three topologies, namely star, tree, and mesh. Depending on the topology, data can be collected, processed, or analyzed within the sensor node, at the local monitoring station, or at the remote data center. Above the link layer, in original version, ZigBee specifications define their own network, security, and application layer protocols. A drawback of the original ZigBee version is that the ZigBee devices need adaptation—they need additional hardware to transfer data to the internet. However, new versions and technologies aim to address this issue. For example, new ZigBee IP specification builds on the IPv6. This version incorporates the new application framework CoAP (Constrained Adaptation Protocol)—an application protocol for energy constrained devices. Similarly, 6LoWPAN has generally implemented a simplified IPv6 protocol above the link layer of the IEEE 802.15.4 standard. Header compression and packet fragmentation reloading is implemented by adding an adaptation layer between the IP layer and the link layer, which is a reliable method to achieve protocol adaptive between IPv6 network and the sensor network [9].

2.2.2 LPWAN technologies

While most of the state-of-the-art environmental monitoring systems use short-range low-power wireless data dissemination networks to transfer information locally (~few hundred meters) from the data acquisition points to the power-unconstrained base stations, and then use another technology (such as e.g., GPS) to further transmit data to greater distances, LPWAN concept brings up a new approach by encompassing at the same time very low power consumption per node and long distance wireless transfer capability. These benefits, however, come at the price of low data rates, somewhat higher delays, and sometimes more expensive infrastructures. As mentioned, three representative technologies that compete in this field are LoRa, Sigfox, and NB-IoT. All of them use narrowband point-to-point transmissions. Sigfox and LoRa use ISM radio spectrum (868–869 MHz in EU and 902–928 MHz in USA) while NB-IoT uses licensed LTE spectrum. Narrower band limits the data rates but it is beneficial in saving energy and it also enables high interference resilience and great number of devices in the network. A limitation of using 868 MHz is the regulative in EU, which allows the transmission duty cycle of only 1%. This will limit the number of messages to 6 packets per hour. However, with slightly higher energy consumption LoRa provides much higher throughput than Sigfox, which allows sending more information in the transmission time slot as compared to the Sigfox. Precisely, LoRa provides the speeds between 300 bps and 50 kbps (depending on the spread factor and the payload), NB-IoT of around 200 kbps, while Sigfox of 100 bps. Sigfox technology enables for the greatest point-to-point range (of maximum 40 km for rural environments) while the shortest range is offered by NB–IoT technology (~10 km). On the other hand NB–IoT can be easily integrated into the existing LTE infrastructure, i.e., no additional base stations are needed. However, LTE uses licensed spectrum, therefore NB–IoT infrastructure (without relying on priory established economically feasible LTE system) would have drastically larger price of implementation as compared to LoRa and Sigfox. Among three technologies, NB–IoT can provide the best quality of service (latency, jitter, and throughput) and the highest scalability (with around 100,000 nodes per cells as compared to the LoRa's and Sigfox's 50 thousand devices per cell). A more thorough analysis on LPWAN technologies is given in Refs. [11,12]. It may be concluded that LoRa and Sigfox will serve as the lower cost technologies with high coverage and very long battery life while NB–IoT is directed to the devices that primarily require high QoS and low latency. Sigfox may be more applicable in systems where data are sent infrequently, and where small throughput, very long battery life, and large coverage are required. On the other hand, a big advantage of NB–IoT systems is that it can be easily added to the existing LTW infrastructure, offering higher QoS as compared to the LoRa and Sigfox. Its deployment practically depends on the will of the mobile network operators to enhance their functionality to IoT applications. Finally, regarding the cost, latency, power consumption, transmission range, reliability, and independency of the solution development, LoRa may fit in the middle between Sigfox and NB–IoT and therefore it may cover a broader range of applications. This application space can be described in the following characteristics of LoRa technology:

(a) It is supported by most of the technology leaders,

(b) It operates at the ISM band and therefore the solutions can be developed independently (in contrast to NB–IoT, which has to be setup within cellular network),

(c) It provides satisfying data rates, higher than Sigfox but lower than NB–IoT, and satisfying ranges (up to 20 km point-to-point),

(d) Among three technologies, only LoRa provides Adaptive Data Rates (ADR) which offers three classes of devices,

(e) Compared to Sigfox, LoRa additionally provides data encryption,

(f) LoRa supports both Timed Difference of Arrival (TDOA) and Received Signal Strength Indication (RSSI) for localization,

(g) NB–IoT devices need regular synchronization, which will consume additional energy while LoRa uses asynchronous ALOHA protocol which enables longer sleep periods and therefore better battery utilization.

All of the aforementioned technologies have their advantages and drawbacks, and with all the advantages in the framework of IoT applications, they still have not been widely deployed so far. To the best of our knowledge, among LPWAN-based applications for remote environmental monitoring, LoRa has been mostly deployed while only few applications rely on Sigfox and NB-IoT. A list of some of the implemented LoRa-based IoT applications focused on different smart-city aspects is given in Ref. [13]. To solve traffic related problems, IoT system in Amsterdam relies on LoRa infrastructure which covers the whole city by using only 10 gateways. Also, Buenos Aires uses LoRa-based system to monitor parameters such as energy consumption, water level in reservoirs, and environmental conditions. Similar systems have been implemented in Zurich, Sao Paolo, Beijing, etc.

2.2.3 Networking topologies

In environmental monitoring, data dissemination may rely on static or on mobile nodes. Generally, from the topological point of view, a network can be considered as static if nodes are static in space or move without changing the network topology. Otherwise, if the node's movement induces changes in data routes, the network is considered as dynamic. For example, even though the systems for ocean, sea, lake, and river quality monitoring can often physically be considered as having a dynamic infrastructure (with more or less minimal nodes' movements in the water), from the networking point of view, these systems are considered as static. On the other hand, as it will be explained in more details in Section 3, air pollution monitoring systems, besides ordinary static WSN (SSN) organization, may contain mobile nodes such as those carried by community/individuals—Community Sensor Networks (CSNs) or installed on vehicles—e.g., VSNs. In these scenarios, in case the data dissemination is to be performed in multi-hop ad-hoc manner (for example by using mobile phones or vehicles as the intermediate nodes to route data from other mobile nodes), the network would be considered as being dynamic. Fully dynamic systems for environmental monitoring, i.e., those that rely on fully dynamic multi-hop mesh topology, to the best of our knowledge, are still not implemented. Instead, VSNs and Community Sensor Networks (CSNs) rely on mobile nodes, but the networking topology and link stability belong to the category of mobile and stabile point-to-point communications, since they use cellular links for direct data transfer to the data monitoring point. As it will be shown in the given examples, combinations of SSNs with CSNs or VSNs are also possible and might sometimes be highly efficient.

Generally, three types of network architectures/topologies have been used in applications for low-cost remote environmental monitoring. In the same manner as in Table 1, in Table 2 these architectures are listed and shortly described.

Typical examples of the N.2 and N.3 networks are depicted in Fig. 3A and B, respectively.

With regards to the categorization given in Table 2, both CSNs and VSNs can be categorized as belonging to the class N.1.

From the given descriptions and discussions, it can be observed that the networking standards and protocols for environmental pollution monitoring are converging in two directions: (a) toward the combination of the short range (IEEE 802.15.4, Bluetooth, WiFi, etc.) technologies with some of

Table 2 A classification of the data dissemination methods—wireless topologies.

Topology class	Description
N.1	A single point to point link between a smart sensor and the data management system. Alternatively, smart sensor can send data to a mobile device (e.g., mobile phone) for further long range wireless transmission
N.2	Star or cluster based (extended star) topologies with multiple sensing nodes and a single sink/gateway node. Sink/gateway node can be further connected either directly to the data center (e.g., via USB port to a server), or it can additionally be equipped with a WWAN module (e.g., 3G/4G transceiver) for sending data to the remote data center
N.3	Multi-hop mesh topology with sensing nodes, intermediate wireless routers, and a sink/gateway node. Here, the sink node can further transmit data to the data center similarly as in N.2

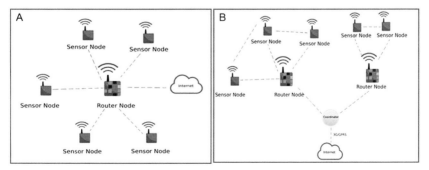

Fig. 3 (A) Typical star-based WSN; (B) Typical mesh-based WSN.

the WWAN technologies, and (b) toward LPWANs. In the first case, the tendency for incorporating the IPv6, security mechanisms, data compression, and CoAP at the higher layers can also be recognized.

2.3 Power consumption and power management issues

Power management/efficiency is one the most crucial issues in WSN. A WSN node typically uses one of the energy storage methods as a primary power supply module. The mostly used sources of energy storages for WSNs today are rechargeable batteries (Lithium Ion or Lithium Ion Polymer) and super-capacitors. Sometimes, the combination of the two has also been used [14]. These resources need recharge and have limited lifetime. Hence, in order to prolong the node's lifetime and minimize the recharge, it is important for the energy to be managed and used as efficiently as possible. A power management system can intelligently manage the batteries to be charged and discharged at separate intervals of time [15]. There are three approaches in prolonging a node's lifetime: (1) the selection of the appropriate low-power hardware components, (2) software settings/adjustments and implementation of energy-aware and energy-saving methodologies and protocols, and (3) utilization of the energy renewable sources.

Among the components that meet the application requirements primarily regarding sensing, processing, and wireless transmission (such as sensors, microcontroller, and transceiver, respectively), the ones with the smaller power consumption should be chosen. For example, IEEE 802.15.4 transceiver enables for smaller power consumption as compared to the 802.11 transceivers. And while the transceiver is usually marked as the most power hungry module of a sensor node, there is a large difference in sensors' power consumption as well. For example, gas sensors typically consume much more power than temperature sensors.

In the framework of software settings and networking protocols adaptations toward smart–node's energy savings and management, four approaches have been mainly applied:

(a) Adaptively lowering the duty cycle to the smallest possible value. This value should still provide the required accuracy and the necessary data delivery timings. This means that the nodes should be put to sleep mode (after sensing, processing, and data transfer) whenever possible, unless the switching is too frequent, because the sleep-to-wake transition is an energy consuming operation.

(b) Implementation of compressed sensing—i.e., the techniques to reconstruct the sparse signal from a few measurements.

(c) Implementation of data fusion—the elimination of the redundant data from processing and transmission modules with as small losses in accuracy and redundancy as possible.

(d) Implementation of the energy-aware MAC and routing protocols with as less deterioration of the network performance indicators (such as delay, throughput, coverage etc.) as possible.

In addition to the aforementioned approaches, today's WSN nodes are usually equipped with the additional energy harvesting modules. These modules either directly supply the sensor node or they recharge the batteries. Basically, there are a few sources of alternative energy used in WSNs, such as: solar, RF, mechanical (piezoelectric, vibration, wind flow, water flow etc.), and those based on thermal gradient. Thorough reviews on energy-harvesting solutions for WSNs are given in Refs. [16,17]. It has been shown that solar photovoltaic conversion provides the highest power density of around $15 \, \text{mW/cm}^2$ but other ways of energy harvesting can also be suitable in some applications. For example, in animal tracking, the vibration energy can be used to recharge the batteries. The hybrid energy-harvesting systems can also be efficient in specific applications. For example, in air pollution monitoring, the combination of the solar panels, temperature converters, and mechanical (wind flow)—based energy sources can be used.

2.4 Taxonomy of the modern environmental monitoring information systems

Although WSNs are converging toward the standardization, their components, architectures, and technologies still need to be specifically chosen and adapted in accordance to the specific application requirements. In practice, due to the different performance priorities (such as cost, accuracy, coverage, flexibility, battery life, etc.), various designs and implementations differ in many aspects. However, the systems' workflow and the aims of the applications remain common to most of the implementations. The information system should work in the inverse feedback manner to eliminate, decrease, or prevent pollution. Besides sensing and transmission, as elaborated in Ref. [18], it is important that data are stored for effective retrieval. The stored data should be visualized but should also be used by prediction algorithms to forecast the place and the time of the (future) pollution. The systems should also provide the (early) warning module. In summary, the modern information system to remote environmental monitoring should perform the following processes:

1. Continuously sample and visualize environmental parameters. Provide the historical view of data from remote (desktop and mobile) platforms.
2. Predict future values of the pollutants.
3. Discover data anomalies, eliminate noise and errors.
4. Based on the measurements (1) and predictions (2), if (actual and future) parameters are within the normal range, repeat from (1)
5. Otherwise:
   a. Turn/keep alarm system on.
   b. By using actuators and appropriate underlying systems (or manually), take action toward adjusting the actual pollution parameters to fit within the normal range, or toward preventing the pollution to happen.
   c. Test for the improvements.
   d. If the results are not satisfying, repeat from (5.a), otherwise repeat from (1).

In addition to the described functionality, the system should enable functions to continuously monitor the state of the system components, e.g., to recognize the malfunctioning of the node, the link instability etc. Also, system should enable the user to change some parameters remotely, e.g., the sampling frequency, etc.

It should be noted that similar workflow is followed by many other smart-water and smart-city applications, where huge number of various sensors are envisioned to be implemented. With the increasing number of sensors, the processing power can become a limiting factor, especially in data analytics. To generate processing speedups, a number of methodologies have been used. The utilization of Gallium Arsenide (GaAs) is one possible solution [19]. Another set of solutions can be oriented in acceleration of classical control-flow systems with various architectural constructs such as, for examples, those presented in Refs. [20,21]. Grid computing is also a methodology widely used so far. Finally, although initially described long ago [22], dataflow computing is now becoming an attractive paradigm for dealing with big data [23].

Taxonomy of the elements and processes of the modern continuous environmental monitoring system is given in Fig. 4.

The proceeding sections review various academic proposals and practical implementation examples of the IoT-based infrastructures on sensing and data dissemination for water and air pollution monitoring. The presentations

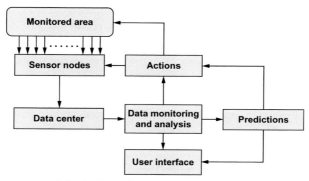

Fig. 4 The taxonomy of the IoT-based environmental monitoring systems.

and discussions are focused on signal processing, data communications, and energy issues. The presented solutions are discussed and compared regarding the criteria/questions such as: what is measured, how it is measured, how data are transferred (the wireless technology, coverage and topology), as well as how the energy consumption is addressed. With the aim to analyze the state of the art in the low-cost WSN-based environmental monitoring applications, the article considers only the peer-review academic sources published in last 10 years.

3. WSN-based WQM systems

Water quality may be influenced by various factors, ranging from the intentional contaminations and deterioration of the distribution system infrastructure to the external sources of contamination, such as urban run-off, industrial disposes, sewage discharge, etc. The contaminated water can have serious affects to human health, aquaculture and to the environment in general. As one of the most serious consequences – annually, there are approximately 250 million cases of water-related disease, with roughly 5–10 million deaths worldwide [24].

The aim of the water quality analyses is to determine the levels of the specific water quality parameters determined by specific standards established for a given purpose (e.g., drinking, aquaculture, agriculture, etc.). After identifying possible pollutants in water, the final goal is to locate and eliminate the sources of contamination as well as to predict the future pollutions in time and space. The analyses include the measurement of biological, physical, and chemical parameters of the water samples.

Traditional WQM procedures are well established and widely used. They have been conducted by taking samples at the regular time intervals, at specific set of locations, and transporting these samples to a laboratory for analysis. By observing the nature of the problem, it can be easily deducted that determination of the actual sampling frequency depends on the importance of the specific sampling station and the expected variability of the water quality data; hence it can hardly be predetermined by standards. For example, the sampling frequency of the traditional drinking water quality measuring procedures is widely accepted to be at least 12 times per year. However, US National Water Information System now collects data every 60 min [25]. There are also other examples in practice where sampling rate should be increased even more or should be made variable. Sometimes, it should be even decreased in order to save the resources. In general, inappropriate sampling frequency leads to inaccurate measurements in time, i.e., to inability to properly handle/catch the signals' variations, which can further lead to the health or environmental related consequences. This inappropriate time and spatial resolution remain the main drawbacks of the traditional measurements. Furthermore, the traditional methods are more prone to errors (especially in data collection and analysis phases), and the whole process is expensive, since it includes significant transportation and human resources.

The actual technological improvements have the potential to address these issues by using new generation ICT infrastructure for continuous near-real time remote monitoring of water quality parameters. The implementation of these systems contributes to timely detection of the deviations in water quality and enables early warnings. From the data acquisition and dissemination points of view, these systems are mainly based on WSNs, primarily because of their fast deployment, low cost, high spatiotemporal resolution, and robustness.

The state-of-the-art in WQM applications have been mainly focused either on monitoring of the (drinking) water distribution systems (WDSs) by installing sensor arrays in pipes, reservoirs, and at the end-user sites; or on environmental (aquaculture) water monitoring, i.e., on detecting and eliminating the water pollution sources in rivers, lakes, and seas. Drinking water distribution systems facilitate to carry portable water from water resources such as reservoirs, rivers, and water tanks to industrial, commercial and residential consumers through complex buried pipe networks [26]. Water from the pipes is then used directly by the consumers. Although raw water or wastewaters undergo treatment process before the distribution, WDS pipelines are easily exposed to intentional or accidental contamination

[27]. Hence, besides monitoring of the reservoirs etc., in-pipe real time WQM is of high importance in pollution detection, localization, and in rapid timely consumer notification. In addition to water quality, water distribution parameters (such as pressure level, leakage detection, etc.) and water usage parameters (such as readings from water meters and control of irrigation) may be monitored as well. Water quality from the rivers and lakes has a considerable importance for the reason that these water resources are generally used for multiple matters such as for: domestic and residential drinking water supplies, agriculture, hydroelectric power plants, transportation and infrastructure, tourism, recreation, and other human or economic ways to use water [28]. Since water from lakes and rivers is often processed and afterwards used as drinking water, the monitoring of lakes and rivers can also be considered in scope of WDSs. Therefore these two geographical entities have been monitored for many purposes. A classification of the research efforts with regards to the deployment location and purpose is given in Fig. 5.

Water monitoring systems installed in seas and oceans are focused on measuring the water pollution affected by the oil leakage, industrial discharges etc. The applications in this area are usually focused on parameters that influence the aquaculture or agriculture, such as chlorophyll, salinity, dissolved oxygen (DO), turbidity, alkalinity, ammonia, nutrient levels, etc. Nevertheless, the system architectures and the monitored parameters are typically similar to those applied in lakes and rivers.

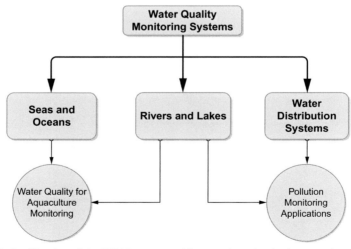

Fig. 5 A classification of the WQM systems with regards to the deployment location and purpose.

Depending on the application area (as given in Fig. 5), a node for water quality sensing can be deployed in one of the two manners (Fig. 6):

(a) A sensors node can be deployed (completely) under the water (e.g., underwater sensors, sensors mounted under the ships, etc.)

(b) Some of its parts (e.g., transceiver) may flow above the water's surface while others (e.g., the probes) may be dived into the water. Typically, these nodes are mounted in buoys.

In the first case, since the communication between nodes–"things" is achieved under the water, the wireless networks for underwater monitoring are usually referred to as Internet of Underwater Things (IoUT). In this case, the acoustic or low RF frequency transceivers are usually used, because they enable for the longer communication ranges (under water) as compared to the microwave RF communications. The second scenario mainly utilizes some of RF-based communication technologies.

Most of the applications use approach (b), where the transceiver and other components (except the probes) are mounted or float above the water surface while the probes are dived directly into the water. On the other hand, for different purposes such as: environmental monitoring, underwater explorations, disaster prevention, military purposes, etc.; the applications for river, lake, sea, and ocean monitoring may utilize both classes of the sensors' deployments. In some of the applications, e.g., in systems for marine environment monitoring, the system design face some additional (electromechanical and data communication) challenges because the nodes are often exposed to the aggressive environment (such as waves, currents, etc.).

Fig. 6 The deployment of the water sensing nodes.

As a common characteristic, the sensor nodes for WQM applications should be designed to be electromechanically robust and the communication protocols should be designed to recover from unstable links regardless of the type of wireless communication.

In accordance to the application area limitations, a number of systems and prototypes have been deployed. Some of them were of generic design, i.e., not restricted to a specific application environment. In scope of this review, some typical examples of the WQM applications will be described with regards to sensing, processing, transmission, and power management technologies. Table 3 gives classification of these examples regarding the application area/location.

An additional list of some of the applied systems for in situ water quality measurement in Korea, Singapore, and US is given in Ref. [46].

From the infrastructure and hardware architecture point of view, the referenced systems vary in sensing objectives and capabilities, the deployment manner, coverage, communication technologies/algorithms, and power consumption/management.

Most of the academic and engineering work has been done toward WSNs' implementations in bigger water surfaces (such as lakes, rivers, and seas). From the academic point of view, this area is more challenging, primarily because: (a) the remote nodes cannot be energy supplied from the power distribution system, which makes them more energy constrained, and (b) the deployment conditions (waves, weather etc.) can affect sensing and communication aspects of the system. On the other hand, to measure the drinking water quality (from water pipes, reservoirs, etc.), nodes can be installed at the specific fixed positions and usually can be power supplied from the power distribution system. This kind of the installing environment makes these systems easier to implement and, consequently, less resource-constraint.

Table 3 Examples of the WSN-based applications for WQM.

Application	References
Drinking water distribution systems (DW)	[29–32]
Rivers, lakes, and seas	[14,24,33–41]
Other (aquaculture, pools, etc.), or not restricted to a specific application area	[42–45]

3.1 Water quality sensing

Strictly speaking, one of the most reliable indicators of water quality is Water Quality Index (WQI). The WQI indicates water contamination degree as a percentage of pure water. The mostly used computational method to calculate this index incorporates physical, chemical, and biologic/organic parameters into a weighted sum, where the weights depend on the importance of each of the parameters for a given application. An example of WQ index extraction based on such approach is given in Ref. [47]. However, in most of the remote real time monitoring applications, a unique WQI has been rarely calculated. Instead, WQI parameters are measured separately. The most common parameters that have been considered are: potential of hydrogen (pH), DO, temperature, conductivity with total dissolved solids (TDS), oxidation reduction potential (ORP), and turbidity. Although, other water parameters have also been measured, such as: oxidation reduction potential [48], chlorine [49], phosphate [50], the presence of *Escherichia coli* (as the main indicator of biological fecal contamination) [51], water spectral properties [31,52], nitrates [14], etc. Some applications use generic indicators of water quality, e.g., optical sensors which measure refractive index as any substance, when dissolved in water, will change the refractive index of the water matrix [31].

In the proceeding of this subsection, main WQ parameters are firstly presented along with the appropriate established reference ranges. Afterward, WQ measurement methods are shortly presented, followed by some examples of the measurement parameters and sensors from literature.

The pH value is defined as the negative logarithm of the hydrogen ion and it measures how acidic the water is. It may vary from 0 to 14, while the value 7 is considered to be neutral. The threshold values for pH are not strictly regulated. For example, the U.S. Environmental Protection Agency (EPA) recommends that the pH values (for drinking water) should be between 6.5 and 8.5, while World Health Organization (WHO) recommends that the pH value for drinking water should preferably be less than 8 [53] . If the values are much lower (below 4), the water is considered to be highly acidic. The acidic water is also corrosive and can cause the leach of metals from pipes. This effect can cause the elevation of toxic metals in water. Consequently, it can be deadly to aquatic organisms, and can seriously affect the human health in long terms. Higher pH value does not impose serious health issues (except for the irritation of mucosa, eyes', and skin).

Although water temperature is not directly related to the water contamination, this parameter is usually measured along with other parameters for two reasons. First, most of the other parameters are temperature dependent, i.e., their interpretation depends on which temperature they are measured. Also, various sensing materials and electronic components are temperature dependent; hence their readings should be calibrated with respect to the given temperature at any time. Second, many studies show that certain temperature ranges provide good environment for the growth of microorganisms. According to the WHO, temperature will also impact on the acceptability of a number of other inorganic constituents and chemical contaminants.

The DO is measured in milligrams per liter (mg/L) and expresses the ability of the water to hold oxygen. Much aquatic life depends on DO. This parameter varies with temperature, pressure, and water salinity, e.g., the higher the temperature, the smaller the dissolved oxygen and vice versa. Therefore, sometimes, DO is measured indirectly, by measuring some of these parameters. However, the water temperature is usually measured along with the DO. According to WHO, the DO value for drinking water should be 5–6 mg/L.

Electrical conductivity expresses the ion concentration in water, while TDS can be used to express the water impurities—it primarily measures the amount of mineral and salt in the water. These two parameters are often correlated, since salt and minerals concentration increase the electrical conductivity. However, the existence of some solids, e.g., sugars, does not significantly affect the electrical conductivity. While being easy to measure, conductivity can be used to indirectly express the water's hardness—the concentration of minerals, mainly Calcium and Magnesium. Also, high conductivity may indicate the presence of industrial waste water. The conductivity is expressed in micro-Siemens per centimeter while the TDS is expressed in parts per million (ppm). Drinking water should have conductivity between 0 and 2500 μS/cm and TDS less than 500 ppm.

ORP measures the oxidizing or reducing potential of water. Any major change in this value can indicate water contamination, especially by industrial wastewater [54].

Turbidity indicates the amount of water clarity, i.e., the amount of particles floating in the water. It is influenced by impurities or algae, planktons, and other microscopic organisms. Therefore, it can also indicate the presence of high bacteria levels and pathogens. The turbidity sensor measures the amount of light that is scattered by the water impurity when the light is directed on a water sample. In drinking water, high turbidity can provide suitable environment for pathogens and therefore can seriously affect the human health. The turbidity is measured in Nephelometric Turbidity Units (NTU). For the drinking water, the turbidity should be no greater than 1 NTU.

In addition to the aforementioned parameters, many other parameters can be measured, depending on the purpose and on the standards for a given region. For example, the Environmental Protection Agency in Ireland describes 101 parameters to determine water quality [33]. Some of them include disinfectants, nitrogen, phosphate, sulfur, micronutrients (such as manganese, iron, cobalt, molybdenum, zinc, copper, boron, selenium, fluorine, and iodine), arsenic, lead, mercury, nickel, chromium, cyanide, silver, aluminum, beryllium, strontium, barium, tin, vanadium, etc.

Chemical compositors of the analyte are recognized by the chemical sensors. Transducers transform the chemical information into a more convenient form such as mechanical, optical, or electrical output. Mechanical sensors make use of physical space that an analyte species takes to detect its presence [55]. These sensors are not used in on–line WQM. On the other hand, optical sensors base their transduction on the interaction of photons with electric structure of the analyte's molecules. They measure variations of optical properties such as color, light absorbance, scattering, luminescence, or fluorescence. Electrochemical transducers are extensively used to measure pH and or DO, but are not limited to these two parameters. They measure parameter concentration and variations by measuring the current flow, or resistance, or potential through the analyte; or by letting the analyte interact with the sensor element and measuring the electrical properties of the sensor's circuit.

A more thorough review on water quality measurement parameters and sensors is given in Refs. [55,56]. It can be observed that, before becoming the integral part of the standardized professional measurement systems, low-cost sensors for multi-parameter remote WQM need to be improved, especially in terms of accuracy and continuous drift correction. However, there is a set of robust and well established commercial low-cost (pH, CO, conductivity, temperature, and turbidity) sensors whose error and some other characteristics are approximately known. For example, temperature and humidity sensors DTH11 and DTH22 have accuracy within the limits of ±2 and $\pm0.5°C$, respectively; and they drain 2.5 and 0.3 mA of current, respectively. As such, these sensors can be used in early/timely warning systems.

The water quality parameters that have been measured in the reviewed examples, along with the corresponding sensor devices are given in columns two and three of Table 4, respectively.

Table 4 The types of sensing parameters and the respective sensors as used in the reviewed researches.

References	Parameters	Sensors
[29]	pH, turbidity, DO	IH2, OBS-3+, Redox sensor
[30]	Conductivity, turbidity, water level, pH	YL-69, LDR and LED, 3in1 pH meter with inverting operating amplifier, probe method
[33]	pH	N/A
[34]	Temperature, brightness	Underwater sensors designed for the purpose of this work at the University of Queensland
[31]	Refractive index	Optiqua EventLab
[24]	pH, conductivity, TDS, DO, turbidity, temperature, water depth	YSI 600XL
[35]	Temperature, pH, conductivity, and DO	DS18B20, Atlas Scientific ENV-40-pH, ENV-40-EC-K0.1, and ENV-40-DO
[36]	pH, DO, conductivity	OrbiSint W CPS, OxyMax W COX 41, ConduMax W CLS12
[14]	Nitrate concentration	Ion Selective Electrodes (ISEs)
[37]	Temperature, pH, and humidity	DS18B20, PI10 Electrode, HIH-4000-001
[38]	Temperature, chlorophyll, turbidity, salinity, conductivity	ISO/IEC/IEEE 21451-2 sea water probe
[39]	Temperature, DO, pH, water level	DS18B20, PH400/450, D-6800, UXI-LY
[40]	pH, conductivity, DO, temperature, chlorophyll, ammonia	Hydrolab Series 5 multi-parameter probe
[41]	Temperature, pH	LE-438 electrode (both temperature and pH) – METLER TELEDO (manufacturer)
[32]	ORP, pH, salinity, water level, turbidity, temperature, flow	Only methods were specified: electrochemical, N/A, conductivity-based, ultrasonic,

Continued

Table 4 The types of sensing parameters and the respective sensors as used in the reviewed researches.—cont'd

References	Parameters	Sensors
		optical, voltage across the diode, N/A, respectively
[42]	Temperature, pH, ORP, conductivity	DS18B20, SEN0169, SEN0165, DFR0300
[43]	Temperature, pH, turbidity	DS18B20, E201-C-9, optical (LED-based designed circuit)
[44]	DO	Only methods were specified: polarographic and (fluorescent-based) optical
[45]	pH, turbidity, temperature	N/A, [57], LM35DZ

Because of the application simplicity and relatively high accuracy of the available sensors, the mostly measured water quality parameters are pH and temperature followed by DO, conductivity, and turbidity. Other parameters, such as chlorophyll, ammonia, nitrate concentration etc., come at third place. Most of the applications are designed to measure just a few parameters. It is mainly because the low-cost measurement systems for remote monitoring are still not tested enough on reliability and accuracy in various environments. As such, they are still mostly implemented in early warning and alarming systems.

3.2 The architectures of the WQM systems

Depending on the application environment and purpose, various hardware architectures have been proposed. The view of the two representative examples of the sensor modules are given in Fig. 7.

The fully integrated devices (which belong to class H.1 from Table 3) represent more compact and elegant plug-and-play solutions, but they are usually more expensive and less flexible for the adaptation to the various scenarios outside the scope of their primary dedication. Moreover, most these solutions are heavy weight and of large physical dimensions. On the other hand, these systems are often well tested with regards to the standards. One such example is Libelium WaspMote [58].

The solutions that belong to the classes H.2 and H.3 are more flexible in the process of adapting and building various scenarios, and can be configured to have as less power consumption as possible, but yet to meet the

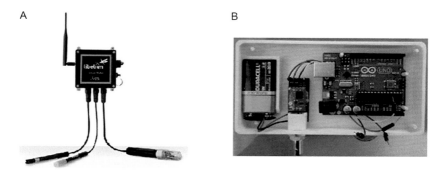

Fig. 7 (A) Typical H.1 node [58]; (B) Typical H.3 node [59].

application requirements. For example, Arduino UNO has generally smaller power consumption as compared to the Raspberry Pi platform. Hence, in the situation when it meets the processing, timing and interfacing requirements of the application, and when the power consumption is a critical parameter, Arduino can be chosen as a node's platform rather than Raspberry Pi. Although more flexible, H.2 and H.3 systems are usually developed and programmed from the scratch, often by using components from different vendors. Instead of focusing on network design and implementation, the developer has to deal with interoperability and compatibility issues, as well as with sensing quality issues such as: calibration, accuracy, testing, etc.

In WSNs, the node's transceiver is the most crucial part of the hardware, since it determines the application coverage, the data rates, and around 70% of the node's power consumption. In the case of WQM systems, small amount of data is sent from the sensor nodes, therefore data rates are not of crucial importance. In the framework of these applications, low power consumption and coverage might be the primary requirements.

Most systems in this application area use radio waves as their transmission medium. When a transceiver is mounted above the water's surface, short range communications (ZigBee, WiFi, etc.), long range communications (LPWLANs, 3G/4G etc.), or combined systems can be used. A study on performance evaluation of an IEEE 802.15.4 WSN on a coastal environment [60] shows that, in case of in-buoy mounted sensor nodes, the most crucial setting is the height at which antenna is mounted. If the antenna is at the appropriate height above the sea level, the application specific disturbances such as waves, fog, and humidity have small influence on the network performances (in terms of throughput, round trip time, received signal strength, and packet loss). Underwater communications belong to a different class of data dissemination techniques. A study on specific technological requirements and challenges in WSN-based applications for marine environment monitoring is presented in Ref. [15]. In contrast to the over-the-air wireless communications, high frequency RF communications are not suitable for under water communications, because of the small underwater transmission range. Alternatively, in this application area, the acoustic signals, low frequency RF communications, and optical communications have been explored. The main advantage of the sound wave communications is that, under the water, the communication range is much greater (up to 10 times) as compared to the usual RF systems for WSNs. On the other hand, the acoustic wave has much slower propagation time ($\sim$1500 m/s) and, because of the much lower transmission frequency, has much smaller transmission rate capability as well. The communication links under water may be more unstable, which can cause the retransmissions and, consequently, longer delays, energy consumption, and low bandwidth utilization. The acoustic signals suffer from multipath, Doppler Effect, environmental acoustic noise and are also influenced by the variability of the conductivity, temperature, pressure, and salinity. As an alternative to the acoustic and RF communications, nodes can also use optical communications. This type of communication enables for the high bandwidth ($\sim$Mbps) and ranges of 10–100 m, but has a main drawback that requires the line of sight between the emitter and receiver, which is often hard to be achieved in WQM applications. A comprehensive study on the Internet of underwater things, the applications, the challenges, along with the proposed channel models for underwater communication is presented in Ref. [61]. A software-defined network for underwater communication systems is presented in Ref. [62].

Given the aforementioned limitations and classifications of the sensing, processing, and communication technologies, it can be observed that practically there is no standard or unique technology that would meet the requirements for all the WQM applications. Instead, hardware design, network architecture and organization, communication manner, etc. are application-specific. Therefore, a wide range of solutions have been proposed. This diversity is exemplified in the proceeding of this subsection, where most of the applications referenced in Table 3 are shortly described and categorized in the context of data acquisition and dissemination techniques.

The applications' descriptions below are given with regards to: (i) the purpose of the application; (ii) which parameters are measured and how;

(iii) nodes' hardware organization and data processing elements, (iv) networking technologies, and (v) energy management. Additionally, the systems are observed from the data utilization point of view. Depending on the number of details given in published papers, most of the proposed systems are commented on their main strengths and drawbacks. The specific sensors used in each of the proceeding examples are given in Table 4, respectively.

In Ref. [29], the authors use the integrated SunSPOT module to sample, measure, and transfer pH, turbidity, and oxygen concentration. In order to adjust the voltage levels, the additional signal conditioning circuits were used. What distinguishes SunSPOT from comparable devices is that it runs a Java Micro Edition Virtual Machine directly on the processor without an operating system [63]. The communication is based on one hop point-to-point 802.15.4 link. The network is consisted of multiple sensor nodes and a base station that form physical star topology. The authors also propose an algorithm for transferring the sensor node to sleep mode in the presence of background process, i.e., they implement an energy conservation method. Moreover, a solar panel (with the voltage regulator) is used to recharge the batteries. As such, the system meets the requirements regarding the power management. The main drawback of this system is its coverage—the 802.15.4 technology enables for the radius of only few tens of meters around the base station.

A system presented in Ref. [30] uses WiFi connection to send values of conductivity, turbidity, water level, and pH toward cloud via access point. The sensing node uses a single chip microcontroller (TI CC3200) with in-built Wi-Fi module and ARM Cortex M4 core for processing, sensor interfacing, and forwarding data to the Ubidots [64] cloud. The IoT software platform has features such as real-time dashboard to analyze data and share through public links. Additionally, the program can send SMS messages whenever the monitoring parameters exceed given thresholds. As such, this system presents a good solution to portable WQM with satisfying data visualization and data utilization capabilities. The main drawback of the system is its short range (WiFi) communication and relatively high power consumption of WiFi technology. Hence, such a system can be mainly used to measure (DW) parameters where sensor node can be easily accessible and the access point can be easily available and reachable.

A system proposed in Ref. [33] uses separated modular subsystems for signal conditioning, processing and data communication. Although the applicability of other sensors is mentioned, the authors use only pH sensor (with the appropriate signal conditioning system). Signal conditioning module is attached to the PIC16F877 microcontroller, which is used for the A/D conversion and signal processing. Data are then sent to the MICAz mote (which is equipped with CC2420 Radio Transceiver) for the IEEE 802.15.4—based wireless transmission. The network topology was supposed to be organized in clusters with the cluster heads positioned in different areas along the river, such as in pump house, residential area, factory, and the farm land. Different sensor nodes are supposed to communicate with the appropriate cluster head by using IEEE 802.15.4 technology (at the distances of up to 70 m). The cluster heads are not necessarily battery supplied and they can send the data either wirelessly or by using wire line links. Besides its low cost, the main strength of the system is that it was built on top of the very low power hardware components. As such, it is expected to be of low power consumption. However, none of the other power saving methods that were given in Section 2.3 were used. Also, besides data visualization, no other data utilization was performed. Based on the given description, it cannot be deduced that the topology itself has been actually implemented. Instead, the multi-hop topology was only proposed while the system was tested only for a simple point-to-point communication.

A very interesting information system to remote monitoring of the water parameters is presented in Ref. [34]. The system is dedicated to monitoring of the underwater luminosity and temperature, information necessary to derive health status of coralline barrier (at Moreton Bay, Brisbane, Australia). However, the system can provide a good generic infrastructure to water pollution monitoring as well. The data dissemination is performed in multi-hop manner. The network is organized in clusters with one gateway node per cluster. After reaching the gateway, the data are forwarded to the base station. The authors develop a modified TDMA protocol in order to achieve more robust and energy-efficient local communication between sensor nodes and the gateway within the same cluster. Sensor nodes rely on Crossbow MPR2400 (MicaZ) processing unit. For local communication (~30 m radius), Chipcon CC2420 was used while MaxStream 2.4GHz Xstream radio modem (of higher power) was used for data transfer from the gateway to the base station. The sensor's sampling frequency was set to 1 Hz but, to conserve energy, data transmission toward gateway was performed every 30 s. The system components, except for the sensors, were mounted into the buoys. The most interesting part of the system might be its power management solution. The energy harvesting module has its own microcontroller to manage the energy. As primary energy source, two rechargeable batteries were used. They are charged through a system based on solar cells. While one battery is in use, another one is getting charged to some value. When a battery gets fully charged, it disconnects from the energy harvesting module. It gets connected and continues to supply the system when another battery's voltage drops below a given threshold etc. Besides, the system uses the sleep mode of the transceiver. The system is designed to report for its "health" as well, i.e., to report the status of the network (connected, not connected, etc.), the status of the battery, the humidity level into the buoy, etc. One drawback of the system could be its small clusters' radius and power-inefficient communication between gateways and the base station.

A multi-parameter water monitoring system presented in Ref. [24] comprises of 20 nodes that transfer the sensed data to the main server by using multi-hop communication. This system is envisioned to measure pH, conductivity, TDS, DO, turbidity, temperature and depth over the area of 5kmx5km. However, the system was tested in much narrower geographical field, where the nodes were positioned at a range of 10 m from each other. The field servers contain two main modules: the water quality sensor module and the wireless sensor node, connected via RS232 communication interface. The smart sensor uses ultra-low-power Texas Instrument microcontroller MSP430F1611 with the CC2420 IEEE 802.15.4-based transceiver. For the purpose of multi-hop communication, the TinyOS-based flooding routing protocol was installed on each sensor node. The sensor nodes consume approximately 40 mA of current in active mode and 20 μA in sleep mode. In order to reduce the data traffic invoked by flooding, and therefore decrease the packet loss, the data compression and averaging are proposed. Each node is supplied from the 12 V battery and alternatively from the solar cell panel. The data are visualized and a warning message service is provided to indicate when the measured value exceeds the threshold value. The field servers, however, are bulky and cannot be considered as cheap. This can be a practical limitation for the potentially wider deployment of the system.

A different approach, based on the optical (refractive) index sensing, is presented in Ref. [31]. For the purpose of sensing and communication, a special Optiqua EventLab device was used. The EventLab is specifically designed to detect any type of chemical contamination in the water at the earliest stage [65]. The system was implemented by Vitens—the largest drinking water company in Netherland. The communication between the sensors and the online central data server was based on GPRS. During a data training period, the system learns to handle the natural dynamics of the water quality, and the threshold values for the event detection are established. An event is triggered when a value exceeds the given threshold. In scope of this work, a system of 44 sensors was implemented and tested. The main strength of this work is data utilization—it has shown high sensitivity to a broad spectrum of contaminations and ability to timely detect major pipe bursts.

A system for WQM applied within Lake of Victoria is described in Ref. [35]. The aim of the application is the automation of the water monitoring to support for the sustainable utilization and management of water resources in the Lake Victoria basin. The system is composed of the data center and two nodes—a sensor node and a gateway node. The sensor node sends data and GPS information toward the gateway module via IEEE 802.15.4 communication. The gateway forwards data through the General Packet Radio Service (GPRS) toward the data center. The main modules of the sensor node are: (a) pH, conductivity, DO, and temperature sensors; (b) the Atlas Scientific circuits for interfacing sensors to the microcontroller, (c) Arduino Mega 2560 board (with the integrated ATmega 2560 microcontroller), (d) EM-506 GPS module, and (e) Digi XBee Pro Series 2 RF transceiver. The main components of the gateway are: (a) Arduino Uno board (with integrated ATmega 328P microcontroller), (b) Seed Studio Arduino GSM/GPRS shield, (c) Digi XBee Pro Series 2 RF transceiver, (d) GSM/GPRS module, and (e) a 1 GB memory card. The memory card is used to store measurements when the GSM link is not available. Regarding the power supply, the system uses polymer lithium-ion battery and two energy saving/boosting methods, namely duty cycling (sleep modes) and solar panel to provide continuous battery charging during day time. The system has shown satisfying functionality over time.

A power efficient non-typical approach to the implementation of the WSN for lake WQM (pH, DO, conductivity) is presented in Ref. [36]. The system applies the cluster-based networking architecture with multiple sink nodes. Sensor nodes implement Sensor Medium Access Protocol (SMAC) [66] over the low-power IEEE 802.15.4 physical layer to organize clusters and to communicate with the appropriate sink node through the cluster head. Sink nodes have two wireless interfaces—besides CC1020 Chipcon IEEE 802.15.4 transceiver, they also pose the Simcom CDMA module to transfer real-time data to the remote data center. Sensor nodes use Atmel AT91R4008 microcontroller (that runs TinuOS-based code) and the same low-power transceiver. All nodes are battery supplied and additionally address power consumption issue in three ways: (a) by using low power hardware and power saving communication protocols, (b) by lowering duty cycle, and (c) by using solar panels. Solar panels are smaller in sensor nodes as compared to those installed in sink nodes. All sensors are put in IP-68 waterproof cases. The system was experimentally tested in the lake of Huguangyang World Geopark for a longer period. The main strength of the system is that it uses all kinds of energy saving methods (mentioned in Section 2.3). Also, based on the derived values, it has shown that it has the potential to replace the conventional pollution detection methods. However, besides simple data visualization, data are not utilized in any other way.

An efficient architecture for river monitoring is proposed in Ref. [14] and implemented/tested at the River Turia, Valencia. Here, 8 motes were installed along the course of the river to measure the temporal and spatial evolution of nitrate concentration. The gateway was placed at the middle of the line formed by 8 sensors (along the river) that cover the end-to-end distance of ca. 35 km. The sampling period was once in a week, justified by the fact that the nitrate concentration does not vary quickly. Motes have incorporated two different energy buffers (battery and super capacitor), an approach to energy conservation (low duty cycle), and an approach of energy harvesting techniques (commercial 54 mm × 42 mm solar panels). The buffers are charged during the day. Once in a week, the M41T94 RTC timer wakes up the nodes to perform the measurements and to send data toward the gateway node. The data transfer is performed on top of the IEEE 802.15.4 standard directly or, if the gateway is not the neighbor of the sender, through the intermediate nodes, in multi-hop manner. The super-capacitor was used as the first energy storage device. A rechargeable battery is used as the secondary buffer, only if the capacitor does not have enough energy to supply for sampling-processing-transmission cycle. The results show that, with the proposed duty cycle, the energy from super-capacitor is enough. Also, all the messages were transferred correctly. Retransmissions occurred in only 3% of the transmitted packets, while none of them needed a third attempt to reach its target. The work is important in the framework of data acquisition and dissemination, because it provides a multi-hop based long range remote monitoring with additional use of very efficient power saving methods.

Another system for sea WQM is proposed in Ref. [37]. Although primarily dedicated to the aquaculture systems, the system composition—architecture is of general purpose. Precisely, each of four sensor nodes, equipped with pH, temperature, and humidity sensors, send data via

GSM module to the base station. Signal processing is performed by using AT89C52 microcontroller. If the pH value exceeds the given thresholds, i.e., if it is greater than 10 or smaller than 4, the warning message would be sent to the farmers, otherwise, data are sent to the workstation for analyses and storage. The system was tested on functionality and feasibility and can be considered as a good laboratory of wide area water monitoring. However, there is no description on power management or power saving methods as well as on data utilization.

The authors in Ref. [38] describe a network of smart sensors based on ISO/IEC/IEEE 21451 suite of standards for both interfacing the (temperature, chlorophyll, turbidity, salinity, and conductivity) sensors, and the appropriate wireless communication devices. Two types of nodes are proposed: (a) a buoy equipped with a SIMCom GPS/GPRS module for long range transmission, and (b) nodes that participate in IEEE 1451.5 compliant network, which relies on Nordic Semiconductor MultiCeiver technology, with the NRF24L01 + PA + LNA module of a maximum output power of +20 dBm. For the processing purposes, the 32-bit ARM Cortex-M4 microcontroller was used. The proposed system is robust, flexible, and scalable. New devices can be easily added without having to make any further changes to the network. However, the authors do not address the energy issues of the nodes nor do test the system on power consumption.

A system proposed in Ref. [41] was tested in an artificial lake. Here, several monitoring areas were covered by clusters of sensor nodes. Smart sensors sense pH and temperature and send data via ZigBee protocol to the respective base stations. Each base station further forwards the measurements to the remote monitoring center via GPRS communications. In this work, the signal processing is performed by the MSP430F1611 Texas Instruments microcontroller. Additional circuits are used for signal conditioning (i.e., amplification and A/D conversion). Each sensor node also contains CC2420 Zigbee transceiver, and since the transceiver is the greatest power consuming element of the node, a special circuit was designed to isolate and to shut down the transceiver when it is not transmitting or receiving data. The base stations additionally contain GPRS modems (with specific MAX232 UART interfacing circuit). Overall, the presented system can be used to cover large monitoring areas. It is flexible, of low cost, and of expected long node's lifetime.

Another applied system to WQM and control for aquaculture based on WSNs is presented in Ref. [39]. The aim of the application is to measure (and control) parameters such as temperature, DO, pH, and water level in near–real time. Although designed to include a greater number of nodes, the system was tested by using only two sensor nodes. The smart sensors collect water quality parameters and send them to the base station through ZigBee technology. Their main components are: probes, signal conditioning and A/D conversion circuits, the ATmega16L microcontroller, and CC2520 ZigBee transceiver. Each sensor was designed and positioned to communicate with the base station via the short range low power IEEE 802.15.4 wireless links. The gateway was directly connected to the data center (host computer). It is composed of only the CC2520 transceiver and the appropriate circuits for RS232 and USB PC interfacing. Instead, the data center (PC) does data processing for the gateway node, i.e., it receives the data from the transmitting nodes and issues commands to control the actuators. Through the incorporated GSM module, it also enables for the data visualization and analysis as well as the alarm system. For software realization, the system uses LabVIEW. The actuators enable for the change of specific parameters, e.g., when DO or pH fall below a given value, the appropriate aerators and alkaline buffers are activated to increase the values. The data were sampled every 3 min and the system was tested in Tanggu district at Tianjin aquaculture for 6 months. Except for the duty cycling and the selection of power efficient hardware, no other means of power efficiency/management were used. The application was tested for the battery performance, sensor data, and network performance and has shown satisfying results. Overall, the system utilizes low power technologies and power management, provides data visualization and some sorts of alert and control.

Another aquaculture monitoring system is presented in Ref. [40]. The system was implemented in three fish farms in Italy. Three underwater nodes were deployed in each cage at the depths of 16, 25, and 40 m. Two nodes were equipped with the probes that measure temperature, conductibility, pH, Redox, DO, depth, chlorophyll, and turbidity; and the third one with the probe that measures temperature, conductibility, pH, DO, salinity, and depth. All three smart sensors were also equipped with the AppliCom SeaModem and the Sapieza University Networking framework for underwater Simulation Emulation (SUNSET SDCS). The SUNSET SDCS solutions are currently commercialized by WSENSE [67] and they enable the implementation of novel protocols. Also, the framework enables the development of the additional modules for easy integration of various probes and implementation of the specific sleep and wake up algorithms. In addition to the three smart sensors, one central node was installed at 6 m of depth and connected to the floating surface buoy. This node was equipped with the

3G/4G router to transfer data to the remote location. Three smart sensors send data through the ultrasound communication to the central node, and the central node further transmits data via 3G/4G link to the data center. The nodes were configured to report the measurement every minute for 30 min. After that period, the nodes went to sleep for 5 h. In addition to the duty cycle reduction, the system uses novel energy-harvesting equipment—a sea-propeller that utilizes the water current energy. The tests have shown that the system can be suitable for long term environmental monitoring. It is also flexible and can be configured for various other applications. Although not enriched with any sort of data utilization, this system is interesting from the infrastructure (for data acquisition and dissemination) point of view.

In Ref. [32], authors measure DW quality, water distribution, and water usage in water supply tanks in an Indian village. Parameters such as ORP, pH, salinity, and turbidity were used as inputs to a logistic regression model in order to extract a simplified and unique water quality index. Three sensor nodes were developed, each consisted of: (a) sensors, (b) Arduino Nano microcontroller, and (c) LoRa FM module. Modules were installed in three different locations and they send information to the LoRa gateway device. The gateway is equipped with a GSM/3G dongle to forward the data to the cloud. In the cloud, the data analytics is performed. Also, from the cloud, alerts are generated and sent to the authorities, and the data are fed into the web page for visualization. The system presents a good example of completely wireless low power and low-cost solution to the water related data analysis and utilization. Similar attempts to implement LoRa technology in WQM are presented in Refs. [42,43,68]. However they are not presented in more details because they lack in some important features such as functionality testing, energy saving, and gateway power autonomy, respectively.

Although widely described in literature, based on the search from various on-line databases, NB-IoT technology has been rarely practically deployed in remote WQM. One NB-IoT based system was presented in Ref. [44] and tested in Fang Shan, Beijing. This system is dedicated to aquaculture dissolved oxygen monitoring and automatic regulation. Two types of sensors were used to measure DO, namely, optical sensor and polarographic sensor. Polarographic DO sensor is cheaper but the service life of this sensor is short, and the sensor needs regular maintenance. Optical DO sensor, on the other hand, has high precision and long operation life. Both sensors are connected to the NB-IoT module via RS485 interface. This module is connected to the PLC control terminal, and the whole system is connected to an aerator. The aerator increases the oxygen level, when needed. Besides, the NB-IoT module provides the connectivity to the remote server, where DO data and the status of aerator are sent every 5 min. The software of the remote system is based on Android mobile operating system. It provides the following services/information to the user: (a) the state of the aerator at any given moment, (b) alarm system (when DO levels exceeds a given threshold), (c) queries on the historical data, and (d) remote manual control function of the aerator, in case of emergency. The system testing shows short response time and significant reduction of the aerator's daily loss and consumption. Although equipped with a solar battery, the system consumes current of 5 Å. It was not tested on the operational lifetime.

As in the case of NB-IoT, although well described in literature, Sigfox-based systems for remote WQM have been rarely deployed. An example of such system is described in Ref. [45]. Even though this system can be used to gather various geo-referenced data for different purposes, in this case the system provides remote and continuous information on water parameters (pH, turbidity, and temperature) along with the measurement geographical position. The main components of the node are: (a) sensors, (b) Arduino MKRFOX1200 board with integrated ATA8520 Sigfox module, (c) GPS module. At the other end of the wireless link, the Sigfox gateway is connected to the Sigfox cloud, which further transfers data to the application server and mobile application. By using low duty cycling, limited number of messages (140 messages/day, as specified by Sigfox), low power data transfer, and additional solar panel, the system is very well suited for long-range power-autonomous operation when the amount of data to be transferred is small. However, in some applications, sampling and data reporting frequency such as the one used in this system might be too low.

Relying on the described approaches and technologies to data acquisition and dissemination, the applications described above are classified in Table 5. The classification is made regarding the following criteria: the hardware composition, the communication technologies, the network topology, the application area, the coverage degree, and the energy management. It is important to specify the criterion used to describe the coverage range. We considered an application to be of small coverage if the distance between sensing device and data visualization/collection/analysis center is of few hundred meters. The application is considered to be of medium coverage range if this distance is in range of a km. Finally, the application is considered to cover large area if the distance between data generator and data monitoring center is of order of more than a few km. In Table 5, the selection of low power hardware and technologies is not mentioned as a specific power conservation method.

Table 5 Examples of the existing solutions classified with respect to hardware architecture, communication technologies, network topologies, application field, coverage, and energy management.

References	Node's HW composition	Comm. technologies	NW/Topology	App area/coverage	Energy saving and management
[29]	H.1	IEEE 802.15.4	N.2	DW/Small	Duty cycle + solar panel
[30]	H.2	WiFi	N.1	DW/Small	N/A
[33]	H.2	IEEE 802.15.4	N.2	River/Medium	Very low power components
[34]	H.3	IEEE 802.15.4 + 2.4 standalone link		Sea/Large	Duty cycle, solar panels, power saving comm. prot.
[24]	H.2	IEEE 802.15.4	N.3	River/Large	Compression Averaging Solar panel
[31]	H.1	GPRS	N.1	DW/Large	N/A
[35]	H.3	IEEE 802.15.4 + GPRS	N.1	Lake/Medium	Duty cycling Solar panel
[36]	H.3	IEEE 802.15.4 + CDMA	N.2	Lake/Large	Power saving comm. prot. Duty cycle Solar panel
[37]	H.3	GSM	N.1	Sea	N/A
[14]	H.2	IEEE 802.15.4	N.2/N.3	River/Large	Super capacitor Duty cycling Solar panel

[38]	H.3	GPS/GPRS IEEE 145115	N.1	Sea/Large	N/A
[39]	H.3	IEEE 802.15.4 + GSM	N.2	Sea (aquaculture)/Large	Duty cycling
[40]	H.2	3G/4G + Sound	N.1	Sea (aquaculture) / Medium–Large	Duty cycling Sea-propeller
[41]	H.3	Zigbee/GPRS	N.2	Lake/Large	Duty cycling
[32]	H.3	LoRa/GSM/3G	N.2	DW/Large	N/A
[42]	H.3	LoRa	N.1	Medium	Solar panel
[43]	H.3	LoRa	N.1	Medium	N/A
[44]	H.3	NB-IoT	N.1	Large	Solar panel, duty cycling
[45]	H.2	Sigfox	N.1/N.2	Large	Solar panel, limited number of messages

As it can be observed, flexible low-cost (H.3 class) solutions constructed from off-the-shelf components have been deployed in most of the cases. And while low-cost data acquisition sensor kits are suitable for high-coverage continuous WQM systems, a disadvantage of using these platforms is that fewer water parameters can be measured. Also, as compared to the traditional measurement methods, these measures are potentially less accurate and impose drift over time.

From Table 5, three main approaches to data dissemination can be observed:

(a) Traditional short range technologies for covering smaller areas. In scope of this framework, IEEE 802.15.4 technology is mainly recommended.

(b) Combined short range traditional technologies with long range technologies for medium to large coverage. In scope of this framework, the combination of ZigBee with GPS/GPRS can provide best performances due to the small energy consumption of the technologies, combined large coverage, satisfying data rates, etc.

(c) LPWAN networks for medium to large coverage, sometimes combined with cellular networks. Among LPWANs, LoRa has shown to cover broader range of applications.

Except for the case where the super-capacitor is used as the primary power supply source, all of the other solutions use Lithium-Ion or Lithium-Polymer batteries for this purpose. Most of the applied systems use duty cycle as the main method to save energy. Solar panels are also often used to provide additional energy source. The exception is one presented case of underwater sensing, where the propeller was used to utilize the water current energy. Finally, compression and data fusion methods for energy conservation have been rarely used in WQM systems. Also, variations of the low-power MAC protocols were rarely explored in these applications. From the application list given in Table 5, the systems employed these two approaches to data conservations were presented in Refs. [24] and [34,66], respectively. Having in mind the environmental influences on wireless link stability, some of the applications (e.g., [35]) have involved memory cards to store measurements when links are not available. Many applications use some sort of alarming/warning system while some of them use additional intelligence to handle the natural dynamics of the water quality [31] or in data prediction purposes. An interesting extension to the basic functionality of the WQM system is the module which provides the information on the status of the battery level, the humidity within the node, the status of the network link etc. One such extension is proposed in Ref. [34]. This kind of system's functionality is of great importance to the systems for remote monitoring, especially in WQMs where the link or node's failures are expected. Finally, some applications [39,44] use actuators to influence the environment in back-propagation manner.

4. WSN-based AQM systems

Air pollution may cause serious damage to human health, especially to people with cardiac and respiratory diseases. Although there is a lack in quantification of long-term-exposure effects to human health, air pollution is widely recognized as one of the main factors that cause stroke and lung cancer. Outdoors, the pollution is mostly caused by road traffic, but other factors influence it as well, such as: industrial activities (e.g., incomplete burning of fuels, oils etc.), human behavior, etc. Many studies on local and (more often) wider geographical areas on air pollution have been conducted. It has been shown that, because the air flow is typically turbulent, the air quality varies over a relatively small scale in space and time. This can be justified by the numerous studies on local air pollution and its spatial distribution, which may sometimes come with somewhat surprising results. For example, in Ref. [69], the authors show that, in addition to the known outdoor pollutants, meteorological conditions (such as wind direction, temperature, humidity, etc.) may significantly influence the local air pollution distribution to the extent of becoming even more important than the location of the main sources of the pollution. In this study, by using linear regression model and Artificial Neural Networks (ANNs) on 1 year collected data from a low-cost sensor node and meteorological station, the authors derive the result/conclusion – that the contribution of the local road traffic on local pollutant concentration is lower as compared to the contribution of the meteorological parameters. The reason for this may rely in the fact that the temperature, wind and humidity variations directly affect the pollutants' dispersion or accumulation locally. Similarly, an interesting conclusion was derived in Ref. [70], where the authors show that the nitrogen dioxide (NO_2) concentration at the vicinity of the (London Heathrow) Airport is mainly influenced by the non-airport air pollution components.

In general, public concern on air pollution has been mostly focused on urban (outdoor) areas. This is why most countries have regulated urban pollution so far. In accordance to the population concerns and interests, most studies on air pollution monitoring have been focused on easy-accessible (user friendly), efficient, reliable, accurate, and timely detection and visualization of the outdoors air pollution levels. On the other hand, various factors may influence air quality (AQ) indoors as well, such as: microbial contaminants, gaseous pollutants (e.g., carbon monoxide, carbon dioxide, etc.), or dust and aerosols. A study described in Ref. [71] shows that cooking, smoking, and spraying pesticides all generate the mass of particles smaller than 2.5 μm ($PM_{2.5}$). The reports of the US Environmental Protection Agency (EPA) are even more concerning; stating that indoor air pollution can be two to five times higher than outdoor contamination. Last estimation of WHO says that some 3.8 million premature deaths are attributable to household air pollution [72]. Finally, people today spend more than 90% of the time inside institutions, homes, etc. Therefore, continuous AQ measurement inside the buildings might be of even greater importance than the urban pollution monitoring.

The aim of the AQ analyses is to determine the concentration levels of the specific air pollutants—i.e., to find the specific pollution values and hotspots that can make possible for the policymaker to react toward the elimination or decreasing of the pollution sources and their influence to the human health. As in the case of WQM, the analyses include measurements of biological, physical, and chemical properties of the air. Historically, approaches for air pollution monitoring generally use expensive, complex, stationary equipments, which limit who collects data, why data are collected, and how data are accessed [73]. Traditional AQM procedures have been conducted by taking samples at the regular time intervals (typically with an hourly resolution), at specific stationary set of locations, and by reading the values directly at the measurement points. They use accurate and reliable sensors but, primarily because of their price (which is typically of few thousands to tens of thousands dollars per equipment), these sensors are sparsely dispersed and therefore cannot meet the coverage requirements. They are also heavy-weight, of large physical dimensions, of high power consumption, and the readings are mostly updated rarely (usually hourly, by using average models). The positions and the density of conventional sensing stations (as they are usually sparsely installed away from the roadsides) do not meet the application requirements as well. By analyzing the wind flow field at an intersection in London urban area, a study in Ref. [74] shows that the pollutant concentrations may vary over a space with a magnitude of few meters, and over time with magnitude of few seconds. Consequently, even though the locations of the conventional sensor nodes aim to be carefully selected, and the temporal resolution might satisfy the requirements of the rural areas, the conventional systems are not able to provide the precise data in larger urban areas, where the parameters of interest follow high spatiotemporal distribution.

An alternative to the conventional (expensive, sparsely deployed) measurement systems is the implementation of the (next generation) densely deployed and low-cost WSNs. WSNs can be placed anywhere as the conventional wired network cannot be deployed, e.g., volatile places like high-temperature areas, chemical and toxins prone areas [75]. When applied in AQM applications, this technology can provide high spatial and temporal resolution. It can acquire data in the geographic scale of a few tens of meters, and can send measurements every few seconds or minutes. The data should be available in near-real-time to the population, academy, authorities, policy makers, etc.

However, as described before, the implementation of the WSNs is not an easy task, because of many technological limitations such as: energy consumption, link instability, the environmental (electromechanical or chemical/biological) influences, etc. Therefore, in accordance to the application requirements, various solutions have been proposed. The applications are mostly divided regarding the location and the range/coverage of implementation. Within this framework, they can be categorized as either dedicated to the indoor or outdoor environments. Regarding this criterion, some representative applications are classified and listed in Table 6.

4.1 Air quality sensing

Depending on the application area, specifically, if the monitoring system is dedicated to the indoor or the outdoor environment, different air parameters may be evaluated. Also, different organizations (such as EPA, European Commission, WHO, etc.) and different countries set their own (different) standards on normal ranges of various AQ parameters as well as on the overall AQ indicator.

While there is no consensus on AQ indexing as well as on the final list of the pollutants that should be monitored to evaluate the AQ, some parameters can be distinguished as being mostly measured by the existing systems for

Table 6 Examples of AP monitoring system prototypes and practical implementations.

Application	References
Systems for outdoor/urban AQM	[76–84]
Systems for indoor AQM	[71,85–91]

AQM. The implemented systems and the academic prototypes for outdoor AQM are mostly focused on measuring: carbon monoxide (CO), nitrogen dioxide (NO_2), ground level ozone (O_3), ammonium (NH_4), PM, sulfur dioxide (SO_2), lead (Pb), temperature and humidity. On the other hand, indoors, the core parameters that are mostly used to describe the overall air conditions and quality are: temperature, humidity, carbon dioxide (CO_2), particulate matter (PM), and Volatile Organic Compounds (VOC).

In order for a single understandable information-value on the air pollution to be provided to the public, from the values of the aforementioned parameters, a single Air Quality Index (AQI) has been usually calculated. Most commonly, the numerical values of six air pollutants (CO, SO_2, PM, O_3, NO_2, and lead) are used to calculate the AQI as the worst index of separately calculated indices for each pollutant. However, different countries use different calculation methodologies and scales to express AQI. For example, US EPA and Singapore categorizes AQ in six categories in scale from 0 to 500. Based on such calculation methodology, the AQI value below 100 is considered to be of good to moderate quality for human health. The AQI of 101–300 is considered to be unhealthy for sensitive groups to unhealthy, while the AQI value of 301–500 is considered to be very poor to hazardous. In EU countries Air Quality Framework Directive specifies AQI in range of 0–100. Other countries use different approaches, e.g., Canada and United Kingdom use 10 point scale value derived from non-linear combination of various AQ factors while Ireland use the same scale but with the AQ Index (AQI) calculated as the worst index of separately calculated indices. A review on various methods for AQI calculation is given in Ref. [92].

In the proceeding of this subsection, a short descriptive review on the aforementioned most commonly measured AQ parameters along with their impact on overall pollution and human health is given. Specification of the AQ measurement methods is then followed by some examples of sensors from literature and practice.

One of the main indicators of the urban air pollution is the PM. The PM is used to express the density of the particles in the air. It is measured in $\mu g/m^3$. A research presented in Ref. [93] has shown that the exposure to the PM_{10} (that indicates the mass of particles between 2.5 and 10 μm) and $PM_{2.5}$ can seriously affect human health. The particles enter through the respiratory system and go to the lungs, where they can cause the inflammation and become toxic to the human body. Long term exposure, even to the small PM levels can cause cancer. Major sources of PM are motor vehicles, residential wood burning, industrial processes, grinding operations etc.

The ozone (O_3) is the reactive gas and powerful oxidant. It is mostly created as a result of chemical reaction between VOCs and Nitric Oxide (NO) in sunlight. In urban cities, the common contributor is automobile traffic exhaust [78]. It may have higher values during the (sunny) summer months. High ozone concentration can severely impact people who suffer from the respiratory disease, especially those who suffer from bronchitis, emphysema, and asthma.

Carbon monoxide (CO) is a colorless, odorless, tasteless, poisonous gas that is produced by the incomplete burning of various fuels, including coal, wood, charcoal, oil, kerosene, propane, and natural gas [85]. Main source of increased CO (up to 95%) is vehicle exhaust. Carbon monoxide levels are typically highest during cold weather, because cold temperatures make combustion less complete and cause inversions that trap pollutants close to the ground [94]. When inhaled, CO enters the bloodstream where it binds to the hemoglobin (which is the main conductor of oxygen through blood). This invokes the reduction of oxygen in blood cells which further reduce the delivery of oxygen to the body organs and tissues. Very high levels of CO are not likely to occur outdoors [95]. The concentration below 2% have been considered as being safe to human health, while extremely high values of CO levels (above 40%), which are possible in enclosed environments, can cause death.

The emissions from motor vehicles or other combustion processes, combined with oxygen in the atmosphere produce NO_2. The NO_2 is mainly present outdoors, due to the motor engine exhaust, and the burning of fossil fuels (coil, oil and gas). However, it can be also produced indoors by unvented heaters and gas stoves [85]. High level or NO_2 and/or long exposure to NO_2 can cause pulmonary edema, emphysema and other bronchial symptoms and disease such as wheezing, coughing, colds, flu and bronchitis.

Sulfur dioxide is an irritant, colorless, reactive gas that is mainly produced when sulfur-containing coils are burned. The main sources of SO_2 pollution are industrial complexes. High level of SO_2 and/or long–term exposure to this gas can also cause respiratory symptoms.

Lead (Pb) is a heavy metal that is emitted from motor engines (which use petrol containing Pb tetraethyl) and from various industrial processes. Most of the Pb particles are small enough to enter the respiratory system and to induce toxicity. Besides respiratory and cardiac disease, Pb exposure can affect the different parts of the human body such as kidneys, liver, nervous system, etc. This element is known as a powerful neurotoxin. Even in small doses, Pb can induce health-related problems to younger population.

Volatile Organic Compounds (VOC) is a set of unhealthy chemicals that vaporize and then can be inhaled by humans and animals. One of the main implications of the VOC is that the compounds react with other chemicals (e.g., Nitrogen Oxides) to form ground-level ozone (O_3). Main outdoor sources of VOC are gasoline vaporization, burning fuel, natural gas etc., while indoors, various aerosols can produce VOC. The harmful effects to VOC's exposure range from eye irritation, headaches, dizziness, visual disorders etc. to lung, liver, kidney, and central nervous system disease.

The CO_2 value is typically used to express the ventilation level in indoor environments and can indicate human occupancy inside a closed space. However, high human occupancy often increases the levels of odor, sweat, etc. Consequently, these, and other bio effluents, can increase the level of the CO_2. Levels over 1000 ppm indicate problem with air pollution or ventilation. Some studies show that health related symptoms can appear even at the levels of 600 ppm. Long term effects of exposure to high levels of CO_2 may lead to serious health-related effects such as chronic inflammation, kidney failure, bone atrophy and loss of brain function [96]. For this reason, CO_2 has been included in many indoor AQM applications.

A descriptive review on air pollutants and their influence to human health is presented in Ref. [97]. Two representative organizations such as WHO and EPA give details on the way the pollutants get in the air, their effects on human health and ecosystem, as well as on the existing standards [72,95].

Professional, highly accurate AQ sensors that represent industrial standards are typically expensive, energy inefficient and are not easily adaptable to the high coverage, distributed, power-autonomous systems for remote and continuous real time monitoring. They also need significant resources for their periodical maintenance and calibration. As compared to these instruments, low-cost sensors still cannot achieve the same accuracy level. However, their accuracy and range typically can satisfy the requirements of approximate measuring. These nodes can be part of the alarm systems for timely warning while they bring other benefits at the same time, such as: low cost, small size and light weight, power autonomy, flexibility, high sampling and reporting frequency, etc. As such, they can also provide real-time information that allow for the timely assessment of the air pollution level as well as the development and the implementation of the pollution prevention methodologies.

There are two main groups of measurement and sensors, namely gas sensors and PM sensors.

Low-cost gas sensors may be: catalytic, optical, thermal, electrochemical, infrared, solid-state (semiconductor), and Surface Acoustic Wave (SAW). Among them, electrochemical semiconductor (MOS) gas sensors have been mainly used, because of their low price, relatively high reliability, high sensitivity, and fast response time. Electrochemical sensors are based on a chemical reaction between gases in the air and the working electrode of an electrochemical cell that is dipped into an electrolyte. In a MOs, also named resistive sensor, semiconductor, gases in the air react on the surface of a semiconductor and exchange electrons modifying its conductance [5]. Their selectivity of the low-cost gas sensor is, however, somewhat low and is affected by air temperature and humidity. This can make the differentiation of some gases more difficult. For example, O_3 readings may interfere with NO_2 readings. A more comprehensive review on gas sensor technologies is given in Ref. [98].

Different methodologies have been used to measure PM as well. Conventional and professional PM sensors usually use either Tapered Element Oscillating Micro-Balance (TEOM) or β-Attenuation Analyzers. On the other hand, although not very accurate, because of their small cost, size, and light weight, low-cost commercial PM sensor nodes are mainly based on Optical Particle Counters (OPC) followed by Nephelometers. Both types of sensors rely on light scatting or light obscuration optical principles. The measurement is based on using a proportionality that relates scattered light to a specific diameter of the PM particle.

An important aspect of low-cost sensor implementation is their calibration. The calibration is mainly performed by using one of the two approaches: (a) by using high-quality (expensive) sensors as the calibration reference point—i.e., the low-cost sensors are exposed to the specific air pollutant concentration (with predefined values), and their parameters are adjusted such that the readings match or are close to the readings from the professional sensors, and (b) by establishing the ground truth value with the estimated reference [76]. But, no matter how precisely the calibration of the sensors is performed; they are sensitive to meteorological conditions and need time to acclimatize when the monitoring environment is changed [99]. Also, the higher the sensitivity, the higher is the interference (in readings) coming from other gases in range. Most manufacturers of the H.1 types of nodes keep the details on the calibration confidential. Also, in literature, little information can be found about the calibration of low-cost commercial sensors. An example of a simple, but still effective calibration method was presented in Ref. [100]. The examined sensor node (A) was placed in a plastic sealed container along with professional-grade AQ sensor module (B). To derive the outputs of the gas sensor in ppm instead of default mV, an equation based on linear model were constructed, with unknown coefficients *a* and *b*. Different gas concentration conditions were created in the controlled excitation environment. Both measurements from A and B were taken. The parameters *a* and *b* were then derived using linear least-square estimation method on the extracted data from A and B. After calibration, the model was applied to the second set of data and the results derived from A and B were compared. The results show good correlation. In Ref. [5] the authors review the calibration techniques for low-cost gas and PM sensors given in literature, and they have derived the following results: (a) regarding CO and NO, multi-linear regression model gives best calibration results; (b) for NO_2 and O_3 sensor calibration, Support Vector Machine (SVM) and Artificial Neural Networks (ANN) are shown to be the most suitable models; and (c) for PM sensor calibration, linear models have shown best results, with Khöler theory being a promising method as well.

The air quality parameters that have been measured in the reviewed examples, along with the corresponding sensor devices are given in columns two and three of Table 7, respectively.

An important drawback of the low-cost AQ sensor systems is that current sensors are not capable to measuring very small particles. However, these particles may pose greater risk to human health [101]. Generally, when temperature and humidity sensors are placed near gas sensors, their readings will significantly differ from the actual values (obtained from the professional instruments), because of the heat produced from the gas sensors. Also, gas sensors need some time to achieve the working temperature, i.e., to adapt to the environmental temperature. This makes the energy conservation by duty cycle harder.

4.2 The architectures of the AQM systems

Smart AQM sensor nodes have similar hardware organization such as those dedicated to WQM, and can again be classified in accordance to the categorization given in Table 1. The respective examples of realization of smart sensors for air pollution monitoring are given in Fig. 8.

Again, fully integrated plug-and-play solutions are relatively easy to implement. Typically, they use calibrated ready-to-use sensors and have robust electro-mechanical design. They are typically equipped with the most representative wireless transceiver technologies for densely deployed

Table 7 The types of AQ sensing parameters and the respective sensors as used in the reviewed researches.

References	Parameters	Sensors
[76]	NO_2, O_3, temperature, humidity, VOC, dust, noise	ELM Perkin Elmer [114]
[77]	Light, temperature, humidity, CO	Hamamatsu S1087, SHT11, MiCS-5525 semiconductor-based sensor
[78]	PM, CO, NO_2, O_3	SGX Sensortech (PPD42NS, SGX-4CO, MICS-2714, PPD42NS)
[79]	PM	Alphasense OPC-N2
[80]	NO_2, CO, temperature, humidity, ambient pressure	Libelium Waspmote Plug & Sense
[81]	PM	GP2Y1010AU0F
[85]	CO_2, CO, Chlorine, O_3, temperature, and humidity	Libelium Gases Pro Board
[86]	Temperature, humidity, light, gas	LM35DZ, HIH-4000, NSL-19M51, TGS 2600
[87]	Temperature, humidity, ammonia (NH3), nitrogen oxide (NO), benzene, and CO_2	Winsen Electronics, LM36, MQ-135 (for NH, NO, benzene, and CO_2)
[88]	(CO, Liquid Petroleum Gas, CO_2), (temperature, humidity)	MQ2, DHT11
[89]	CO, CO_2, $PM_{2.5}$, PM_{10}	TGS2442, TGS4161, DustTrack DRX Aerosol 8533
[90]	Temperature-humidity, CO, CO_2, light	SHT10, MQ7, T6615, LDR5
[71]	$PM_{2.5}$, VOCs, humidity, temperature	Dylos DC1700, AppliedSensor iAQ-engine, Sensirion SHT15, Sensirion SHT15
[91]	CO_2, CO, NO_2, SO_2, O_3, Cl_2	BME282, INE20-CO2P-NCVSP, 4-CO-500, 4-NO2-20, 4-SO2-20, OX-A431, 4-Cl2-50
[82]	AQI (smoke)	MQ-135
[83]	PM	Alphasense OPC-N2, Plantower PMS5003, Plantower PMS7003, Honeywell HPMA115SO
[84]	PM, temperature-humidity	PMS5005, SHT20

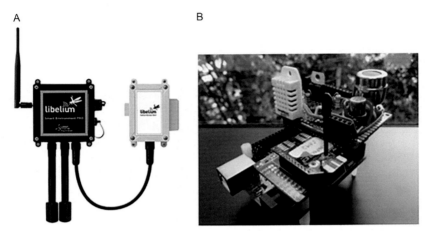

Fig. 8 (A) Libelium smart environment PRO [100], (B) A node designed in Ref. [102].

WSNs such as ZigBee. However, these solutions lack in the design flexibility, weight, size, and cost. A study on H.1 class of sensor nodes for AQM is presented in Ref. [103]. According to the predefined criteria (that were based on Rank Based and Compensation Method), among 15 integrated sensor nodes, the calculated weighted sum indicates that, at the time of comparison, Libelium Waspmote Plug & Sense shows best overall performances in urban environment monitoring. One such node equipped with 16 sensors is presented in Fig. 8A. An example of a H.3 low-cost node was developed and presented in Ref. [102], Fig. 8B.

While being similar in many technical aspects, within the design space of physical robustness, maintenance costs, wireless transmission requirements and network organization, AQM systems may pose some specific characteristics as compared to the smart sensors for WQM. The WQM nodes may be deployed at the surface of the seas, oceans or underwater. Consequently, they may more likely fail in operating. On the other hand, AQM nodes are less affected by the environment. While WQM nodes are usually deployed in remote locations, AQM nodes can be relatively easily accessed and repaired/replaced. Another difference between AQM and WQM nodes is the power source requirement. Most of the AQ systems are mounted on objects with the available power supply from the power distribution systems, large vehicle batteries, or easy accessible rechargeable systems, e.g., on traffic or city lights, on vehicles, or mobile phones, respectively. On the other hand, WQM systems require usually the full power autonomy of the sensor node.

Regarding the wireless communications and networking, besides relying on static wireless links, AQM systems may also rely on mobile nodes, i.e., they may be organized in form of CSNs or VSNs (Figs. 9 and 10, respectively). A comprehensive review and comparison of the SSN, CSN, and VSN wireless architectures and technologies for AQM is presented in [104].

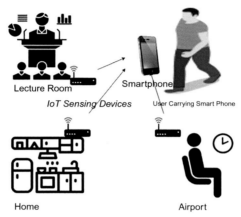

Fig. 9 Community sensor networks.

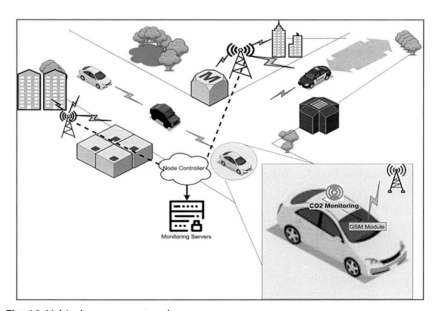

Fig. 10 Vehicular sensor networks.

Static WSNs (SSNs) for AQM are typically mounted on the traffic lights or streetlights. As such, they do not suffer from mobility issues (such as link instability), size/weight and power consumption limitations. These systems can be of high accuracy and reliability as well. However, the position of the nodes is not always optimal and the maintenance implies that the professionals have to visit the measurement points—stations in order to repair or replace the nodes.

The CSN nodes are envisioned to be carried by people. Although most of the up-to-date implementations have been conducted by a number of volunteers carrying specific equipments for data acquisition, cell phone-based solutions are more likely to become standard for CSNs in the future. As such, these systems would be cost-efficient, of high coverage, and would directly provide the information on air quality to the user. Moreover, smartphone-based systems, besides of directly providing the information to the user, can be used for other purposes as well. This concept, which is characterized by the citizens' (or general public) involvement in data collection for further use, analysis and evaluation, is now recognized as participatory sensing or anthropocentric opportunistic citizen sensing [105,106]. Sometimes, it is also called crowd sourcing—a concept that describes voluntary participation of individuals and organizations to collect data for further compilation and analyses by third party. This way a large volume of data on everyday human activities can be stored, processed, and fused, in order to be used afterward for various other purposes as well. In the context of AQM, the main advantage of these systems is that they directly measure and visualize the pollution level at the user's vicinity, and can also warn a user based on data collected from elsewhere. Moreover, the system can be connected to the specific user personal health data, which would enable the alarm systems to be adjusted accordingly. On the other hand these systems have their drawbacks as well. For example, since they are handled by the user, the number of sensors, their weight and size are limited. Also, the user movement can induce artifact in signal acquisition, which can further influence the system's accuracy and reliability. A study on reliability of available fixed and smart-mobile solutions for AQM is presented in Ref. [107]. In this study, both passive tubes and electrochemical (Azimut station) gas sensors were used to measure NO_2. Nine passive tubes were installed at 3 m altitude on street pillars by using protection cases. Two of them were placed near the local AQM center, to check for the sensors' accuracy. On the other hand, Azimut mobile stations were carried along the given trajectories during 2 weeks at the human level by volunteers who, at the peak traffic hours, passed near each of the installed passive tubes. While passive tube collected data for latter analyses, the Azimut station has sent data in real time. However, by comparing the results taken from the local AQ station with those taken from the fixed stations and those taken from the mobile sensors, the results have shown that readings from passive tubes are more close to those obtained from the AQ station. More importantly, the readings from the mobile nodes have shown significant differences with the readings from fixed nodes. This can increase the awareness on various additional factors that influence the air pollution monitoring procedures. Power consumption may also be a concern in CSNs, but the batteries can easily be recharged, since the users have direct access to the power supply system. Another drawback of the CSNs is their random deployment. Precisely, nodes are mobile and their spatial distribution can lead to the redundant data at some areas while missing data at some other geographical areas. Because of the given constraints, the CSNs are rarely deployed as the network of standalone devices to sense and transfer data to the data monitoring center. Instead, in order to make the system as accurate (and reliable) as possible, the most common implementation of the CSNs combine them with other types of networks, e.g., utilize the SSNs for sensing and (short range) wireless data transfer to the mobile phones, from where data are further disseminated via cellular networks.

A good example of a CSN, based on a specifically designed device, is AirQuality Egg [108]. From the manufacturer's specifications, the Egg is a WiFi-enabled device that uses sensors to record changes in the levels of NO_2, CO_2, CO, O_3, SO_2, and VOC. Each Egg can detect at least one air contaminant. It tracks any changes and automatically uploads the data to the cloud where it can be accessed through web portal, mobile app, or by manual download by connecting the Egg to a computer [108]. To the best of our knowledge, the only available standalone phone-based (CSN) solution to the AQM is provided by BROAD Life mobile phone interface. Its accuracy is analyzed in Ref. [109]. The authors show that this phone may present the appropriate AQM tool only when the pollutants' levels are high enough. Under the controlled experimental conditions, it has been shown that the phone's response was linear only at higher particle concentration. This qualifies it as potentially suitable only in polluted environments, where it can serve as an alarm system to the users. Also, due to the energy constraints, the phone provides measurements only on a press of a button. Therefore it cannot be used as a tool for continuous monitoring.

Finally, VSNs have a great potential to be used in comprehensive and efficient urban environment monitoring. So far, they have been mainly deployed to measure traffic congestion [110], parking statistics [111], road conditions [112], etc. Recently, there is a growing interest in VSNs for pollution monitoring. They are typically carried by the public transportation systems, e.g., trams, buses, trains, taxies, bikes, etc. As such, the nodes can have stabile power supply and are not constrained in energy, since they can be supplied from the vehicle's power source. They can be equipped with many sensors, and their size and weight do not present an implementation constraint. Also, these systems are of high and dynamic coverage with easily accessible and easy-to-maintain sensor nodes. On the other hand, similarly as in CSNs, their dynamic spatial distribution may not be optimal at specific points in time, since vehicles may experience traffic jam and/or, at a given point of time, they can coincide to be grouped at specific geographic areas, which would impose the imbalanced data reporting—such that it is redundant at some areas and sparse/missing at other areas. Also, noise in data can be introduced by the vehicle system, i.e., the vehicle will typically produce additional pollution which is added to the environmental pollution. Besides, when sensor nodes are mounted on the roof of the vehicles, protecting them from the meteorological conditions becomes a challenge. On the other hand, by adding extra protection, the air flow might be disturbed which may lead to the additional measurement inaccuracies.

As in case of CSNs, besides completely VSN-based solutions, with the aim to utilize the advantages of different data transfer techniques, combination of data dissemination methods may be used as well. One such approach is given in Ref. [113]. The authors propose the combination of the short range SSN communications with long range VSNs. Precisely, SSN clusters are envisioned to be deployed over different parts of the city. They use ZigBee communication to transfer data with the time and location stamp toward the sink node, where the data would be stored in local memory. On the other hand, when mobile nodes (which would be mounted on public transportation objects) would pass in the vicinity of the sink nodes, they would automatically collect stored data and upload them via fifth generation (5G) or Long-Term Evolution (LTE) systems to the data center. Because of the low complexity, the nodes' hardware of these systems may be of low cost. They utilize power efficiency of the cluster-based ZigBee network infrastructure as well as the fast and reliable 5G/LTE data transfer. Power supply (and consequently power consumption) is not a limiting issue as well, because it can be adapted from the vehicle power system.

Given the aforementioned limitations and classifications of the sensing, processing, and communication technologies, it can be observed that practically there is no standard or unique technology that would meet the requirements for all the AQM applications. Instead, hardware design, network architecture and organization, communication manner, etc. are application-specific. Therefore, a wide range of solutions has been proposed. This diversity is exemplified in the proceeding of this subsection, where most of the applications referenced in Table 6 are shortly described and categorized in the context of data acquisition and dissemination techniques.

Once again, but now in the context of AQM, the applications' descriptions below are given with regards to: (i) the purpose of the application; (ii) which parameters are measured and how; (iii) nodes' hardware organization and data processing elements, (iv) networking technologies, and (v) energy management. Additionally, the systems are observed from the data utilization point of view. Depending on the number of details given in published papers, most of the proposed systems are commented on their main strengths and drawbacks. The specific measured parameters and sensors used in each of the proceeding examples are given in Table 7, respectively.

In order to identify and address various issues of using WSNs to monitor urban air quality, 20 ELM modules [114] were deployed in York, UK [76]. The sensor modules have integrated sensing, processing, and GSM communication capabilities. Data on NO_2, O_3, temperature, humidity, VOC, dust and noise are uploaded to the server every 20s via GSM module. When GSM is not available, data are temporally stored in local memory and are uploaded when GSM becomes available again. The work is important because it identifies some crucial problems related to the WSNs for urban AQM such as: variations in sensors' readings, data aggregation, the sensors' positioning, and calibration. The energy consumption was not addressed, but that was not in the scope of the project, because sensor modules are power by AC and mounted on the lamp-posts.

A system for continuous air pollution and metrological data monitoring in urban area (of Taipei) is described in Ref. [77]. Its purpose is to remotely visualize and analyze obtained values of light intensity, temperature, humidity, and CO. The system is composed of two main parts, namely front-end automatic monitoring system and control center. Front-end uses short range ZigBee WSN infrastructure for data acquisition and dissemination toward the gateway node. The gateway transmits data via GSM link to the control center. At the data center, LabVIEW [115] was used to visualize the data, but also to store data and to integrate the storage into the database for further analytics. The smart sensor employs Texas Instruments MSP430F1611 architecture with 12-bit A/D converter, 10KB of RAM, 48KB flash memory. Nine sensor nodes were installed to the traffic single pole to collect air pollution and metrological data every 10 min. The results indicate the main pollution points in the Taipei, along with the time intervals of the peak pollution which is related to the direction of the traffic flow. The results can be also monitored from smart phones, PDAs etc. Hence, both municipality authorities and citizens can benefit from the proposed system. Gas sensors are the greatest energy consumers in these systems. To conserve the energy, CO sensor was switched on and off. However, sometimes, depending on the on/off intervals, this can raise a question on the sensor's accuracy/reliability, since low-cost gas sensors generally need some time to "warm up." The system was not tested on power consumption but it is not practically limited because the nodes are power supplied from the power distribution system. An important issue addressed in this work was the sensor's calibration. The calibration equation was derived by using linear regression on data from sensors and those obtained from EPA monitoring station. The aforementioned characteristics make the system successful and appropriate in many technical aspects. We notice, however, that the equipments-nodes are bulky and hence not easily and likely to be widely implemented.

A system for urban pollution monitoring in the city of Uppsala is presented in Ref. [78]. Three sensor nodes, two coordinators and gateway were positioned in the city center with a close proximity to the country operated AQM stations. Once per hour, each node measure and send data regarding: NO_2, PM, CO, O_3, temperature, and humidity. The network is organized in tree topology with one coordinator for one smart sensor, and another coordinator for another two smart sensors. Both coordinators individually communicate with the gateway. The communication between smart sensors and the gateway is achieved via XBee 868 MHz links. The sensor nodes use Arduino platform (based on ATMega328P), while gateway node uses pcDuino board [116]. It is connected to the Internet router through Ethernet technology. For data storage, analyses, and visualization, the TerraView cloud platform was used [117]. Google Map was used for analyses on correlation between traffic data and the actual readings from the sensors. The overall reliability of the system has shown to be satisfying, as well as the accuracy (as compared to the data provided by Stockholm and Uppsala Country Air Quality Center). However, packet loss delivery ratio is around 80% and the number of duplicate packets is significant as well. The latter phenomena are explained as coming from the fact that two coordinators are in use and in range, i.e., this architecture suffers from collision effect. The nodes were left active all the time (because of the MOS type of sensors), hence energy consumption and nodes' life time remain as the main drawbacks of the system. The main strength of this paper might rely on the fact that it utilizes full multi-hop architecture, which can make the system more suitable where low energy and wide coverage are important parameters at the same time.

A VSN system for collection of dense spatiotemporal data of various phenomena in urban areas is presented in Ref. [79]. A sensing platform was mounted on top of garbage trucks and implemented for 8 months in Cambridge, MA, USA. Equipped with thermal cameras, WiFi scanners, accelerometers, GPS, air quality, temperature and humidity sensors, the presented system can perform various measurements (such as air pollution monitoring and prediction, crowd and station mapping, street driving behavior, identification of the activity patterns and noise, etc.). For data dissemination, WiFi hotspots were used, because it was found to be cheaper as compared to the cellular networks. It, however, does not allow real time data transfer. Furthermore, data can be influenced by the emissions from the vehicle on which it is deployed, although, as found in Ref. [118], the particles emitted from the trash truck are of sizes around 100 nm, while this work (with optical particle counter Alphasense OPC-N2) can sense only particles of sizes between 0.38 and 17.8 µm. Nonetheless, data are still subject to various sources of noise. Finally, we find this project interesting because: (a) it covers the large city area, (b) it can be densely deployed and hence provide high spatiotemporal resolution, (c) with 8-month deployment, a great number of data was obtained to identify air pollution hot spots.

Another example of the VSN for urban pollution monitoring is presented in Ref. [80]. The work is a part of the larger (GreenIoT) project developed in Uppsala, Sweeden. As an attachment to the existing stationary nodes, the work introduces mobile sensors mounted on the roof of the city buses. Libelium Waspmote Plug & Sense nodes are used to measure NO_2, CO, temperature, humidity, and pressure. Stationary nodes utilize 6LoWPAN to transfer data to a gateway. Data are then further transferred by using WiFi. On the other hand, mobile sensors are not energy-constrained because they use bus-offered continuous power supply. Hence, it is not necessary for their wireless data communication technology to be power efficient. The bus itself is powered by a diesel engine and a set of Ultracapacitors. Ultracapacitors utilize the mechanical energy from the brakes. The nodes installed on the vehicles use 4G technology to transfer data to the HTTP API in the GreenIoT cloud. In the cloud, data analytics is performed by using Python scripts.

Two suitable bus routes were selected to cover the largest area in Uppsala. The data collected are displayed on Google maps visualizing both the pollution values and the actual position/coverage of the sensors. The application also has modules to inspect the system's health and memory card to store data when the links are not available. The conducted tests have shown that results for CO are satisfying, while for NO_2 Libelium mote has low sensitivity. Also, 70% of the data are successfully received on the first try. The rest of 30% were stored on the memory card and received afterwards where the wireless link was re-established.

A large scale air pollution monitoring system is presented in Ref. [81]. The authors have developed a system to measure the PM in different measuring points in Bucharest. The main components of the system are: (a) smart sensors, (b) coordinators, (c) the gateway modules, and (d) the server. The appropriate ADC circuit was used to digitize the signal that will be further transmitted out of the sensor module. Data from the sensor module were sent to the IEEE 802.15.4 module. At the upper layers, the network uses JenNet protocol—the enhanced 6LoWPAN protocol for mesh networks. The data are sent to the coordinator, which is connected to the e-Box 4300 module (gateway) via Universal Asynchronous Receiver/Transceiver (UART) interface. The e-Box provides TCP/IP connectivity with the network control center (Apache 2.44 server) where data are stored in a database. The system has shown to be efficient and flexible for integration in the Smart City concept and is important as one of the rare systems that actually incorporated 6LoWPAN in a greater coverage area for AQM. Also, the implementation costs of this system are low. The system did not implement any of the energy saving/management procedures or algorithms.

A laboratory on point-to-point ZigBee system for indoor AQM is presented in Ref. [85]. The sensing node is based on Libelium Waspmote while the gateway is based on Raspberry Pi 2 platform (with integrated ARMv7 processor). The system enables the evaluation of CO, CO_2, ozone, clorine, temperature, and humidity in a school in Doha Qatar. For data visualization, an IoT platform based on Emoncms [119] open-source web application was developed. The system was evaluated only regarding its functionality. No energy evaluation and energy saving methods were used. The main strength of this work is that it demonstrates the almost step-by-step development of a complete IoT-based platform for real time indoor AQM based on the integrated available hardware (Libelium) and software (IoT/Emoncms) solutions.

Another simple representative lab-design of a system for indoor AQM, now constructed from the off-the-shelf components, is presented in Ref. [86]. It is focused on measuring temperature, humidity, light, and some typical indoor air contaminants (such as hydrogen, ethanol, isobutane, and CO). The network is organized in star topology and utilizes Zigbee communication protocol stack with few sensor nodes and one coordinator as its central point. Specific signal conditioning module has been designed for adapting the signals to the range of the other modules. The coordinator is the central point of the network connected (via USB port) to the data center Labview control program, which provides GUI for data visualization and analysis. The measurements were taken and tested under the controlled laboratory conditions, by performing specific excitations on each of the measured parameters, e.g., by increasing the room temperature, etc. The system has some obvious drawbacks, such as: small application area, i.e., not designed for online real time access, does not address the power consumption, etc. However, it represents a good example/guide to a laboratory on systems for indoor AQM.

A system for indoor AQM is presented in Ref. [87]. The target measurement parameters are temperature, humidity, ammonia (NH3), nitrogen oxide (NO), benzene, and CO_2 gases. In contrast to most of the similar applications, it uses multi-hop communication manner for data dissemination. The ATtiny85 microcontroller was used for data processing and XBee series 2 modules was used for communication among the sensor nodes. The energy consumption is addressed by adapting the sampling frequency and transmission period. Normally, the data are sensed and stored. The average values are transmitted every 15 min. However, if the sensed data are alarming (e.g., high CO value or temperature), they are immediately sent toward the sink node. The system is of low cost. The systems is therefore interesting in many aspects—it uses multi-hop architecture in indoor environment, it addresses the power consumption, and is of low cost. However, although tested on energy consumption and the results promise low power consumption, the nodes' lifetime is not clearly presented.

Yet another (rare) multi-hop application for indoor AQM is presented in Ref. [88]. The proposed system measures and visualizes graphically CO, LPG, CO_2, temperature, and humidity values. The ZigBee network is consisted of sensor nodes, a router and a coordinator. Arduino Uno board (with Atmel ATmega328P microcontroller) was used for signal processing. During the testing (in the University building) one sensor was placed in one room, the router in another, while the coordinator and the data center were placed in the third room. To conserve the energy, a special electronic circuit was constructed to put the board and sensors in sleep mode after taking a sample and transferring it to the data center. The functionality test was performed under the controlled excitations. The results were satisfying in terms of system response. Moreover, energy consumption and node's lifetime were measured. As compared to most of the existing solutions, the average power consumption was low, draining about 11 mA. For the given duty cycle, the node's lifetime was also satisfying. However, the derived values from sensors were not compared to those from standard industrial sensors nor calibrated, and therefore the accuracy might be questionable.

A good example of a multi-functional system for real time AQI calculation (and control) inside a building is presented in Ref. [89]. The sensing module is Libelium Waspmote which incorporates the ATmega1281 microcontroller and Digi Xbee communication module to measure and transmit CO, CO_2, $PM_{2.5}$, and PM_{10} values. Five nodes were placed at different indoor areas (conference room, reception, cubicle, cafeteria, and laboratory), with a central node set as a base station and ZigBee coordinator. Each location was mapped to mote's MAC ID. The user can graphically see the actual and historical data on the measured parameters, as well as mean, maximum, and average values of each parameter. Moreover, an ontology-based middleware was developed. It enables the applications to subscribe for the interested context parameter. Also the system is coupled with the Heating Ventilation and Air Conditioning (HVAC) system, to enable the improvement op AQ based on the derived measurements. To obtain AQI values, the authors use EPA calculation formula [120]. The measurements were performed under the different room's occupancy conditions. Additional excitation (e.g., candle burning) was also included to test the sensor's sensitivity. Various functionalities of this system, makes it a good effort/proposal of the IoT-based information system for AQM and control. The authors do not mention any of the energy saving or energy management techniques used in the proposed system.

In Ref. [90], the authors monitor air temperature, moistness, CO, CO_2, and glow in indoor conditions. The star/tree networking topology has been used for data dissemination. The network is composed of two wireless end-nodes, and a gateway. The sensor nodes were equipped with multiple sensors, processing Arduino MEGA (based on Atmel AVR microcontroller) unit, and a wireless module. From the sensor nodes, data are sent via IEEE 802.15.4 wireless links toward the gateway. The gateway (with incorporated Wemos Electronics Mini D1 processing unit) uses only wireless technologies for communication with the nodes and with the Internet. It is equipped both with power amplified XBee version—XBee-PRO module (to extend the communication range, as compared to the ordinary XBee modules), and with a WiFi module. The first one provides the communication between the sensor node and the gateway while the second one enables further data transmission to the MySQL database and Internet web portal. The end user can access data via the web portal built in PHP. The data are visualized both in form of numerical values and in a chart form. The data history is also available. The system provides both desktop and mobile accessibility. Besides the aforementioned functionality, the application incorporates the alert manager, i.e., it informs the user when a specific parameter reaches some abnormal value. The system was tested regarding its functionality and power consumption. The tests have shown that it allows for easy data access via web portal or from a Smartphone. However, according to the performed tests, small power consumption has not been proved (with only ∼3 days of node's lifetime). However, in indoor environments, we do not consider power consumption as a high priority issue, since power can be assured from the AC distribution system. We do consider the proposed system to be interesting in the framework of data acquisition, dissemination, and visualization because it contains the most important parts of the indoor AQM systems and gives priority to the most important issues as well, such as: it uses sufficient wireless transmitter power to cover the area of interest, various networking topologies can be constructed (by using ZigBee protocol), it uses stabile and high-throughput WiFi to upload the data to the database, it uses both desktop and mobile access to the data, and it provides the alert system.

An intelligent home-based sensing system for indoor AQ analytics— AirSense is presented in Ref. [71]. The AirSense is envisioned to automatically detect the pollution events and sources, to classify them, estimate personal exposure and give suggestions in improving the AQ. The sensing platform which measures $PM_{2.5}$, VOCs, humidity and temperature is based on Arduino Uno Ethernet board [121]. The PM sensor was directly connected to the laptop, where manufacturer's software [122] was utilized to capture data. Other sensors were included on-board. Data were uploaded via WiFi link to the cloud server that, besides storing data, runs an analytic engine. The main strength of the work is contained in the methodologies for data utilization. The analytical engine provides pollution detection, identification of the pollution sources and events, and their forecasting in near future. Also, it generates suggestions related to the identified pollutants, and sends the recommendations to the user's Smartphone. The experiments were conducted at two homes for 10 weeks and three other homes for 9 weeks. The proposed system was able to successfully (with an average accuracy of 95.8%) identify three most common household activities that generate significant indoor air pollution such as smoking, cooking and spraying pesticide, as well as their combination. Moreover, it can also forecast future pollutions with an average error of less than 8.1% within 5 min after occurrences of pollution. However, the paper is mainly focused on data analytics. The issues such as node's autonomy (and lifetime) and the sensing/communication coverage were not addressed.

A wider coverage infrastructure for remote AQM was presented in Ref. [91]. Sensor nodes sense temperature, humidity, and six gases (CO_2, CO, SO_2, NO_2, and Cl_2), and send them to their local gateways via XBee modules. Gateway devices are based on Raspberry Pi2 board. The board use Ethernet connection to the Internet. Again, to extend the communication range, both gateway and smart sensors use XBee PRO modules. The sensing platform is based on Libelium Waspmote with the attached sensor interface Gas Pro Sensor Board and a set of calibrated sensors attached to it. As in Ref. [85], the authors use Emoncms IoT application. The main strengths of this system/work for indoor AQM are: (a) its coverage, (b) its accuracy and reliability, (c) it assesses the data loss under different scenarios—relative position of the node and the gateway, (d) it relies on real IoT software platform for data visualization, and (e) great number of measured AQ parameters. The functionality tests were performed for a month in University building. Interesting correlations were noted from the extracted from data, such as the correlation between CO and NO_2. Also, the level of CO_2 depended mainly on the students' presence in the rooms. The system was not tested on power consumption of the nodes.

Similarly as in the chapter on WQM, before closing this subsection with integral table, the proceeding text covers the examples of AQM systems which implement LPWAN. One of them is presented in Ref. [82]. It is dedicated to measure and calculate the AP index by using only one sensor (MQ-135), as this sensor is expected to be sensitive to ammonia, sulfides, and benzene vapors. The main modules of the system are: the sensor node, cloud database, and PC client application. In the framework of technical specifications, the authors use fully IoT-oriented approach based on recent technologies, which is also the main strength of the work. For acquiring data from sensor and transmitting them to the remote location, it uses pycom expansion board and Sipy (Sigfox) module [123], respectively. As mentioned before, Sigfox technology might have the lowest power consumption among the LPWAN technologies and the pycom hardware is designed for low power consumption as well. Hence, the choice of this hardware will make sensor nodes to operate at a very low-power mode, and will make the application of high geographical coverage. In compliance with the hardware design, the software architecture is also based on IoT modules; precisely, the IBM Waston IoT Platform and Google Cloud database were used. The application enables user to view API values in time, the detailed history of the API, and the location of the pollution in map. Although the presented system is very interesting in the framework of data acquisition and dissemination technologies, it has some drawbacks such as: (a) the positions of the sensors displayed in the map are fixed, i.e., no GPS was used to update the information if a sensor moves, (b) in order to conserve the energy, it sends data only 48 times a day (which has impact on time resolution), and it uses only one sensor (which has impact on accuracy), (c) except for providing various kind of visual data presentation, it does not conduct any of data utilization methods.

An experimental study on the capability of AQ low-cost devices to capture spatiotemporal variations of PM and on the ability of LoRa technology to transfer data reliably in long distances (12 km) in Southampton (UK) is presented in Ref. [83]. Six devices, each with four different PM sensors (selected based on their popularity in the literature, their ease of use and availability), have been designed and deployed across two school sites (within the city). One school is located next to a road, i.e., the measurements were influenced by traffic. The closest Southampton Center Monitoring Station (SCMS) was located 1 km away from school A and 2 km away from school B. The results derived from developed sensors were compared to those received from the SCMS. Sensors were attached to the Raspberry Pi board, equipped with LoRa transceiver. The system was powered via Power over Ethernet (PoE) standard. Compared to the values obtained from the SCMS, all of the deployed sensors have shown good results in terms of root mean square error. Among them, however, Plantower PMS5003 and PMS7003 were significantly better than others. The developed system is important because it shows the low-cost sensor's capability to measure PM and the advantages of LoRa networks in remote high coverage monitoring. However, the system has some significant limitations, such as: (a) it is bulky, (b) it is not power autonomous, and therefore it is not clear why LoRa technology was chosen for data dissemination, since LoRa's main advantages are coverage and low power operation, (c) the work covers only a fraction of an AQM system, i.e., it is mainly focused on sensors and a simple communication infrastructure, without any form of data utilization.

An interesting approach to AQM is given in Ref. [84]. Five sensor nodes were deployed over an area of approximate radius of 3 km to extract the AQI from $PM_{2.5}$, PM_{10}, CO, SO_2, NO_2, and O_3. For each of the pollutants, specific index is calculated. Then, AQI is derived as the maximum of the derived values. The system's architecture is composed of three main layers: (a) Sensor layer (monitoring nodes), (b) network layer (access point), and (c) application layer (data processing server, storage server, web server, and mobile application). The STM32F103RC microcontroller unit (with the ARM 32-bit Cortex-M3 microcontroller) implements sampling and adaptive duty cycle adjustment. By default, the node is woken every 10 min to send and transmit data. However, if the difference between two successive readings is larger than a given threshold, the duty cycle is lowered so as to monitor the AQ in time. The transmission manner, which complies with the LPWAN requirements, might be the most interesting part of this work. To transmit data, 802.15.4k baseband signal processing procedures are programmed in microcontroller. The GFSK modulated signal is then transmitted through the TI CC1125 very low power chip at the frequency of 433 MHz and data rate of 200Kbps. The sensor nodes are mounted on the street lights infrastructure. On the other end of the link, the access point (AP) uses general purpose processor to implement open source Software Defined Radio (SDR). The antenna of the AP was installed on the roof of a 15-floor building at the center of the university campus. The proposed system enables the adjustment of the transmit power of the nodes in accordance to their distance to the APs. This significantly saves power as compared to the strictly defined transmission power of the RF transmitter. Moreover, to save energy, adaptive duty cycle is applied along with the solar panel. The sensed data are sent to the cloud database and analyzed in the IoT cloud. The web interface and mobile application enable the visualization of the results as well as viewing of the historical data and data prediction. Given the above description, it can be noticed that the developed system can be a good example of the low power optimized systems for AQM and prediction. The performance of the system is evaluated in terms of: (a) the carrier-to-interference measurement and (b) the accuracy of the derived AQI compared to the reference values.

Relying on the described classification methodologies, in Table 8, the AQM applications involved in the review are classified regarding the following criteria: the hardware composition, the communication technologies, the network topology, the application area, and the energy management.

Similarly as in the case of WQM applications, in AQM systems, due to the greater design flexibility, the H.3 type of node's hardware composition/design is mostly used. However, most of the AQM applications are less constrained in energy because smart sensors are often installed in the locations where some mean of stabile power supply exists (street lights, vehicles, etc.) or are a part of the accessible rechargeable devices (such as smart phones).

Table 8 Examples of some of the existing solutions classified with respect to hardware architecture, communication technologies, network topologies, application field, coverage, and energy management.

References	Node's HW composition	Comm. technologies	NW/topology	App area	Energy saving and management
[76]	H.1	GSM	N.1	Urban	N/A (AC powered)
[77]	H.2	ZigBee/GSM	N.3	Urban	Duty cycle, ON/OFF control
[78]	H.3	XBee/Ethernet	N.3	Urban	N/A
[79]	H.2	WiFi	N.1	Urban	N/A
[80]	H.1	4G	N.1	Urban	N/A
[81]	H.2	ZigBee/ Ethernet	N.2	Urban	N/A
[85]	H.1	ZigBee	N.1	Indoor	N/A
[86]	H.3	XBee	N.2	Indoor	N/A
[87]	H.3	XBee	N.3	Indoor	Duty cycle, ON/OFF control
[88]	H.3	ZigBee	N.3	Indoor	Duty cycle, ON/OFF control
[89]	H.2	ZigBee	N.2	Indoor	N/A
[90]	H.3	ZigBee/WiFi	N.2	Indoor	Duty cycle, ON/OFF control
[71]	H.2/H.3	WiFi	N.1	Indoor	N/A
[91]	H.1/H.2	ZigBee	N.2	Indoor	N/A
[82]	H.2	Sigfox	N.1	Urban	Ultra low power technology, duty cycle
[83]	H.3	LoRa	N.1	Urban	N/A (PoE powered)
[84]	H.3	802.15.4k	N.2	Urban	Duty cycle, solar panel

As such, their wireless transceivers are not very much power constrained. However, in order to use resources as efficiently as possible, and to design AQM nodes that can be moved as required and used anywhere, the tendency of making them as power efficient as possible remains. Within this framework, most of the applications use ZigBee for short range communication, coupled with power amplified XBee versions or GSM for longer range communications, or they aim to utilize LPWAN technologies to cover both ranges. Among the LPWAN technologies, it is noticeable that, so far, LoRa has been implemented more broadly. As it can be noticed from Table 8, in addition to the low power hardware, except for the variable duty cycle and solar-based energy harvesting methods, no other means of power conservations have been used. However, we claim that more efforts should be done toward energy saving and management issues of the power-autonomous AQM node's design, primarily because gas sensors are very power-consuming elements.

Similarly as in the WQM applications, some of the AQM applications (e.g., [76]) have involved memory cards to store measurements when links are not available. Many applications use some sort of alarming/warning system. Most of them use some sort of IoT platform for data visualization and storage, such as for example IBM Watson IoT Platform and Google cloud database. Some of them use intelligent platforms to forecast pollution in near future and to identify air-related activities [71]. Finally, some applications use actuators to influence the environment in back-propagation manner. One such example is given in Ref. [89].

5. Discussion: Observations, challenges and future improvements

The growth of the world's population and human activities has influenced the growth in urban run-off, industrial disposes, road traffic, heating derivatives (such as incomplete burning), etc. The existing (conventional) systems for water and air pollutants' tracking and reporting are not able to properly follow the actual pollution trends and distribution. Their drawbacks can be summarized in: high cost per equipment, low spatial sensing resolution, non-real time data reporting, often non-optimal geographical position of the measurement points (e.g., dependent on the accessibility to the public power distribution system), high maintenance costs, and low degree of data utilization.

The WSNs, which are based on low-cost smart sensors, are recognized to have a great potential in overcoming the technological gaps of the existing systems. However, a number of issues have slowed down their implementation in this field, such as: sensors' configuration and accuracy, nodes' power consumption and management, wireless coverage and communication issues, scalability, security, and environmental influences.

In this article, WSN technology is analyzed in the context of its potential to provide low cost reliable and accurate real time remote air and water quality monitoring. The actual technological solutions are described and classified with regards to the different technological categories and requirements. Most of the innovations in the areas of WSNs and IoT have been created using the hybridization methodology [124]. In order to exemplify the variety of the actual solutions and the difficulties of the WSN implementation in environmental monitoring, some recent prototypes and implemented systems along with their advantages and drawbacks are shortly presented.

From the presented academic proposals and the structure of the in-situ applied systems, it can be recognized that, a modern IoT-based information system for environmental monitoring should contain:

(a) Sensor modules.

(b) Communication infrastructure.

(c) Data storage, visualization, and alarming resources and capabilities.

(d) Data analytics module.

Sensor module is a critical part of the system, since it is involved in in situ transformation of physical parameters into the electrical signals and data, and into transferring these data to remote locations. Moreover, this module is expected to be power-autonomous, physically small enough and of low cost. From the material presented in this article, it can be deduced that both WQM and AQM nodes follow similar node's hardware issues and challenges. One of the main challenges is the accuracy of (low-cost commercial) sensors and their reliability. EPA indicates that many commercially available sensors have not been challenged rigorously under ambient conditions, including both typical concentrations and environmental factors [125]. Accordingly, current sensors should be comprehensively evaluated to ensure that their characteristics (accuracy, sensitivity, etc.) are as close to the standardized instruments as possible. A thorough analysis on performances of the most popular PM sensors is presented in Ref. [83].

Although it is important for as accurate as possible sensors to be implemented, for many of the monitoring objectives, it is not critical to have sensors that meet uncertainty requirements of larger more robust monitors [73], but to:

(1) Know their uncertainty and other performance specifications;

(2) Be able to reference them to the more robust monitors;

(3) Deploy large number so that confidence in the measurement is improved due to many measurements rather than a few.

Most of the sensors are sensitive to various operating conditions such as temperature, humidity etc. Also, their accuracy is influenced by aging, environmental interference, and as well as the low sensor's selectivity [76]. On the other hand, the improvements in selectivity will often decrease sensor's sensitivity. These issues are to be addresses in the context of the composition and the structure of the sensing materials. In order for the various aforementioned factors to be included into the sensor's measurement process, the sensors should be calibrated. The calibration means finding a specific pattern that correlates the measured values from the specific (low-cost commercial) sensor with those obtained from the professional highly accurate instruments. In this article, some of the calibration methods were presented and referenced. The best calibration methodologies for various types of sensors were accentuated.

Energy consumption is one of the main constraints in the WSNs design space. As identified in this article, there are three main roads in prolonging a node's lifetime: (1) the selection of the appropriate low-power hardware components, (2) software settings/adjustments and implementation of energy-aware and energy-saving methodologies and protocols, and (3) utilization of the energy renewable sources.

In both WQM and AQM systems, the sampling and reporting rates are relatively low, which enables the utilization of low power microcontrollers and (low data rate) low-power wireless transmission technologies. Because of their modularity, flexibility, easiness of integration and development, it can be observed that most of the applications use some of the user-friendly platforms-boards such as Arduino Uno and Nano, although at the price of somewhat higher power consumption, Arduino Mega and Raspberry Pi platforms have been used as well. However, results from the researches on power-efficient microcontrollers (such as the one presented in Ref. [126]) indicate that, because of very low power consumption and still satisfying processing capabilities, PIC (8-bit) and the MSP430 microcontrollers might be among the best candidates for interfacing and in-node processing of the environmental parameters. Within the framework of low power wireless data transfer, it can be observed that, if the required transmission distances are small, IEEE 802.15.4-compliant transceivers followed by Bluetooth BTE and

6LoWPAN ones would be the best candidates. On the other hand, in semi–mobile applications (such as VSNs for AQM) or long range applications, the combination of IEEE 802.15.4 with GSM or LPWAN technologies, respectively, would fit within the application requirement space, especially when this space is bounded by the node's power consumption, network coverage, and network capacity. LPWAN technologies overcome the somewhat high energy consumption of the traditional WWAN transceivers that have been used in remote sensing, although GSM technology is fairly power efficient. However, all of the three representative LPWAN technologies (SigFox, LoRa, and NB–IoT) come with some limitations. For example, Sigfox, as the lower power consumption technology, enables only the limited number of the uploading messages and cannot be efficiently used when communication from data center to the node is required. Also, both Sigfox and NB–IoT are operated by infrastructure companies—Sigfox operates on the basis of subscriptions (i.e., the usage of Sigfox's cloud APIs is required), while NB–IoT operates in scope of LTE networking infrastructure. Although there is a set of wireless applications that would be better addressed by using Sigfox or NB–IoT, based on the technical specifications and on the number of applications given in literature, we evaluate that LoRa has the potential to meet the greater number of the requirements for high–coverage environmental applications. Besides LPWANs, the emerging 5G technology promises to become one of the networking infrastructures for IoT applications. The main benefits from this technology are expected to be contained in its minimal latency and very high bandwidth. However, the latency and bandwidth are not of crucial importance to the applications for environmental monitoring. Furthermore, power consumption of the 5G enabled nodes is still not evaluated. Therefore, from the technical point of view, other technologies might better meet the requirements of the environmental monitoring applications. However, this technology is still in its rollout phase. The utilization of this technology in this application area will directly depend on its penetration level and its cost-effectiveness. So far, regarding the environmental monitoring applications, only some frameworks (such as the one given in Ref. [127]) are described.

Since most of the measurement parameters are expected to change slowly over time, the applications for remote environmental monitoring can utilize the software-triggered power conservation methods. Putting the devices periodically in power sleep or deep sleep modes contributes greatly in saving energy. The actual power efficient transceivers consume around 10 mA in transmit mode and current of order of ~μA in sleep mode. The appropriate boards and circuits consume much more energy though.

Therefore, it is important to turn them off when a sensor node is not taking samples, processing, or transmitting. Moreover, the adaptable duty cycle (which depends on the dynamics/variability of the measured parameters) can be used. Implementation of such power-saving method was implemented in Refs. [84,87]. Also, in saving the gateway's energy, the applications would benefit from data fusion, compressive sensing, and from energy aware MAC protocols. However, although widely explored in literature, these techniques have been rarely applied in systems for remote environmental monitoring.

The nature of the network deployment in WQM and AQM allows the utilization of the energy-harvesting techniques. For example, both types of applications can benefit from mechanical converters (to convert water flow or air flow into the electrical energy). On the other hand, the urban pollution monitoring systems and buoy-mounted WQM sensors can both additionally utilize solar energy, while indoor AQMs may rely on temperature gradient converters. As described in this article, solar photovoltaic conversion provides the highest power density. However, the additional solar charge controller device has to be engaged, to reduce the produced voltage and store the energy safely in battery. The functionality of the system may be additionally extended to real time monitoring of the battery charge level, to protect battery from over-charging. Combinations of a few energy-harvesting techniques may additionally increase the nodes' lifetime. Some efficient combinations were presented and discussed in this article.

From the presented discussion on data sensing, processing and dissemination, the consistency, accuracy, and durability of the sensing elements along with the reliability of the wireless links may be recognized as the main factors that slow down the wider deployment of the low-cost WSN-based applications for environmental monitoring. A possible solution that would bridge the gap between the drawbacks of the conventional systems and those based on the WSNs might incorporate both systems, precisely:

(a) In order to overcome the main drawbacks of the conventional systems, WSN-based solutions would enable for much greater spatial resolution and timely reporting. These systems could be primarily used for early warning of the end users and authorities, relatively accurate data presentation, and data prediction.

(b) In order to overcome the main drawbacks of the WSN-based solution (such as accuracy and reliability), the conventional methods of environmental monitoring, which encompass greater number of environmental parameters (e.g., bacteriological parameters) and still perform measurements more accurately, should still be performed periodically as well.

Besides the presented infrastructure for data acquisition and dissemination, as noted before, modern information systems for environmental data acquisition, monitoring, and prediction should incorporate additional modules for data storage, error detection, visualization, alarming systems, and data analytics. Accordingly, in the field of AQM, EPA is evaluating lower-cost sensors for criteria pollutants (NO_2, O_3, PM, and VOC) in collaboration with sensor developers and federal and state partners and is developing data visualization methods, to support its Geospatial Measurements of Air Pollution Program, that could potentially be used to visualize mobile sensor data [128]. The data generated by IoT devices, sensors, and IT supplementary devices are large, so it is necessary to use cloud or fog computing platforms to store and process them. As it has been shown in this article, most of the applications use cloud computing for this purpose. However, since sending all the data to cloud platforms can be expensive or can generate delays, the use of fog computing platforms has grown, becoming more relevant in recent years [13].

The data should be utilized in several domains such as: (a) visualization, (b) alert generating, and (c) prediction. Via web services, mobile applications, and SMSs, visualization and emergency system must utilize data from the IoT infrastructure to timely inform the citizens, authorities, and academy on actual and past data, as well as on the data that need urgent attention. Also, the prediction systems should enable the timely prediction of the pollution incidents, e.g., based on the variation of some other parameters, the system may predict that CO is going to increase to a large value within a specific time interval. Moreover, the intelligent systems should be used to detect anomalies and errors. For example, if the spike appears in visualized data, it is important to recognize the nature of the spike, i.e., if it comes from the parameter variations or from the environmental noise. Since there is a lack of studies in understanding anomalies in environmental data, it is still not clear what contextual information is relevant or should be utilized for the detection of abnormalities of a particular parameter [76]. According to a review given in Ref. [13], IoT research works started to include machine learning or data mining techniques to establish data analytics process just a couple of years ago. The same study finds that only 35% of the publications consider the incorporation of data science approaches (mainly decision tree, neural networks, and time series), and only 15% considers aspects related to the decision support systems. Therefore, as the main pollution sources now might be considered as recognized, as well as main factors that influence the pollution distribution, we consider that more studies should be performed toward the implementation of data mining algorithms on data acquired from both AQM and WQM sensors and environmental data. This would utilize the high resolution data and would be very helpful in the field of emergency response. An example of using Support Vector Machine (SVM) in anomalies detection over the simulated environmental data is presented in Ref. [54].

Finally, although a very important issue, only a few projects indicate to use security mechanism.

From the presented material—academic work and experiences, we deduce that future WQM information systems should provide the integrated solution for continuous, secure and timely monitoring and visualization of various parameters measured in different water categories such as drinking water, environmental, and wastewater. Moreover, the system should provide water consumption information and water waste detection to the authorities and policy makers. To the residential users and authorities it should provide the early warning system and system to improve safety from flooding and water risks. Finally, such system should be able to make predictions and to identify link errors, node failures, etc.

On the other hand, high quality mobile air pollution sensors coupled with Internet technology will greatly expand the amount of information that EPA, states, industry, and the public will have to understand, reduce, and prevent air pollution [73]. Similarly as WQMs, the AQM systems should be integrated into the smart city projects with more or less same capabilities as WQMs but with some, aforementioned, specificities in sensing and wireless transmission methodologies. In order for the greater coverage to be achieved, CNSs and VNSs should be utilized more extensively. In this context, the involvement of mobile network operators would be highly beneficial. Additionally, indoor AQM systems should be more available and utilized.

Relying on the presented work, in the area of WQM and AQM, some possibilities on the improvement of the existing information systems for remote and continuous environmental monitoring can be identified:

(1) The sensing capacities (i.e., number of the measured parameters) of the existing low-cost WQM and AQM systems should be increased.

Comment: This becomes a challenge because it will increase the nodes' power consumption, which may influence the main goals of the system. To overcome the energy constraint, an alternative solution might rely on combining sensing data from different nodes.

(2) Utilization of multiple energy conservation and energy harvesting techniques, whenever possible.

Comment: Solar panels have proven to have the greatest power density. However, alternative methods (such as wind, water flow, temperature gradient, etc.) can be used depending on the implementation environment. As presented and discussed in this article, the combination of various energy harvesting methods can be highly efficient.

(3) Utilization of the algorithms for dynamic adaptive sampling and data transmission rate.

Comment: This approach enables for the sleep/active rate to be adapted to the dynamics of the measured parameters. Having in mind the relatively slow dynamics of most of the environmental parameters, this methodology can greatly contribute to the power conservation of the nodes.

(4) In case of underwater sensors or densely deployed nodes on the water surface, implementation of the wireless multi-hop power aware routing algorithms with data aggregation and or compressed sensing should be considered.

Comment: Data aggregation and compressed sensing should however be used carefully, because they may disable the possibility of recovering the real measurement on each sampling interval.

(5) Exploration and possibly wider implementation of the LPWAN technologies for fixed systems, and VSNs and CSNs for semi-mobile and mobile systems, respectively.

Comment: These networks have great potential, but their accuracy can be influenced by various factors. The collected data are subject to systemic and stochastic noise that is introduced by the sensor, the vehicle system, or mobility patterns [79].

(6) Integration of the storage capacities (e.g., memory cards) into the smart sensors, in order to store data whenever wireless links are not available.

Comment: this is especially important in WQM, where wireless links are expected to be more influenced by the environment. However, as exemplified in this article (in the case of 4G link etc.), from this extension, the AQM applications can benefit as well.

(7) Implementation of the data mining and time series techniques for WQM and AQM classification, event detection, and prediction.

Comment: These techniques are especially important in the framework of data anomaly detection, early pollution detection, pollution prevention, avoidance or elimination.

(8) Involvement of the Geographic Information System (GIS) for temporal and spatial data representation.

Comment: Besides benefiting on data visualization, temporal and
spatial data merged with the data from sensor nodes may greatly help the
data mining algorithms to improve their accuracy and produce new
knowledge on the way the pollution is produced and disseminated
on the different geographical backgrounds and urban environments.

(9) Development and implementation of the user friendly mobile applications for data visualization.

Comment: The data on pollution level should be easy accessible to
the citizens, authorities, and to the academy.

(10) The implementation of the error detection and node's health
reporting system.

Comment: Based on the given criteria (e.g., node is not sending
data for a given period of time or is sending data that are recognized
by the intelligent system as anomalies), the system should be able to
recognize the node's failure, erratic reporting, etc. In some situations,
as presented in this article, it is possible for the additional sensors (e.g.,
humidity sensor) to be installed inside the sensor node, to report for
the sensor's mechanical compatibility, e.g., if fluids have entered
the sensor node.

(11) The systems should implement security mechanisms.

Comment: Although the implementation of very restrictive security mechanisms in computationally-constrained sensor nodes is a
challenging issue, some available mechanisms should be implemented.
This is especially important in drinking WQM systems, where data
can be intentionally corrupted; in which case it could greatly impact
the human health.

(12) When WQM and AQM systems get integrated into the greater
smart-water and smart-city applications, respectively, the data generated by IoT devices, sensors, and IT supplementary devices will be
large, so it is necessary to use cloud or fog computing platforms to store
and process them. In order to process such a big amount of data, some
specific processing paradigms (such as GaAs-based architectures,
dataflow computing, grid computing, etc.) might be used.

Comment: The latter solutions to this problem were listed and
referenced in this article.

References

[1] G. Mois, S. Folea, T. Sanislav, Analysis of three IoT-based wireless sensors for environmental monitoring, IEEE Trans. Instrum. Meas. 66 (8) (2017) 2056–2064.

[2] L. Lombardo, S. Corbellini, M. Parvis, A. Elsayed, E. Angelini, S. Grassini, Wireless sensor network for distributed environmental monitoring, IEEE Trans. Instrum. Meas. 67 (5) (2018) 1214–1222.

[3] K.S. Adu-Manu, C. Tapparello, W. Heinzelman, Water monitoring using wireless sensor networks: current trends and future research directions, ACM Trans. Sensor Netw. 13 (1) (2017). Article 4.

[4] M.T. Lazarescu, Design of a WSN platform for long-term environmental monitoring for IoT applications, IEEE J. Emerg. Sel. Top. Circuits Syst. 2 (1) (2013) 45–54.

[5] F. Karagulian, M. Gerboles, M. Barbiere, A. Kotsev, F. Lagler, A. Borowiak, Review of Sensors for Air Quality Monitoring, EUR 29826 EN, Publications Office of the European Union, Luxembourg, 2019.

[6] A.M. Rangel, T. Sharpe, F. Musau, G. McGill, Field evaluation of low-cost indoor air quality monitor to quantify exposure to pollutants in residential environment, J. Sens. Sens. Syst. 7 (2018) 373–388.

[7] Libelium Plug and Sense Models, 2018. Available online: http://www.libelium.com/products/plug-sense. (accessed on February, 2020).

[8] Memsic, 2020. Available online: http://www.memsic.com/userfiles/files/Datasheets/WSN/micaz_datasheet-t.pdf. (accessed on February, 2020).

[9] S. Yinbiao, et al., Internet of Things: Wireless Sensor Networks, White Paper, IEC, Geneva, Switzerland, 2014.

[10] G. Mois, T. Sanislav, S.C. Folea, A cyber-physical system for environmental monitoring, IEEE Trans. Instrum. Meas. 65 (6) (2016) 1463–1471.

[11] K. Mekki, E. Bajic, F. Chaxel, F. Meyer, A comparative study of LPWAN technologies for large-scale IoT deployment, ICT Express 5 (1) (2019) 1–7.

[12] R.S. Sihna, Y. Wei, S.-H. Hwang, A survey on LPWA technology: LoRa and NB-IoT, ICT Express 3 (1) (2017) 14–21.

[13] R.O. Adrade, S.G. Yoo, A comprehensive study of the use of LoRa in the development of smart cities, Appl. Sci. 9 (2019).

[14] J.V. Capella, A. Bonastre, R. Ors, M. Peris, In line river monitoring of nitrate concentration by means of wireless sensor networks with energy harvesting, Sens. Actuators B 177 (2013) 419–427.

[15] G. Xu, W. Shen, X. Wang, Application of wireless sensor networks in marine environment monitoring: a survey, Sensors 14 (9) (2014) 16932–16954.

[16] K.S. Adu-Manu, N. Adam, C. Tapparello, H. Ayatollahi, W. Heinzelman, Energy-harvesting wireless sensor networks (EH-WSNs): a review, ACM Trans. Sens. Netw. 14 (2) (2018) 1–50.

[17] S.O. Olatinwo, T. Joubert, Energy efficient solutions in wireless sensor systems for water quality monitoring: a review, IEEE Sens. J. 19 (5) (2019) 1596–1625.

[18] P. Knezevic, B. Radnovic, N. Nikolic, T. Jovanovic, D. Milanov, et al., The architecture of the Obelix-an improved internet search engine, in: Proceedings of the 33rd Annual Hawaii International Conference on System Sciences 2000, HI, USA, 2000.

[19] V. Milutinovic, Surviving the Design of a 200MHz RISC Microprocessor, IEEE Computer Society Press, California, LA, 1997.

[20] A. Milenkovic, V. Milutinovic, Cache injection: a novel technique for tolerating memory latency in bus-based SMPs, in: Proceedings of European Conference on Parallel Processing, Municsh, Germany, 2000, pp. 558–566.

[21] A. Grujic, M. Tomasevic, V. Milutinovic, A simulation study of hardware-oriented DSM approaches, IEEE Parallel Distrib. Technol. Syst. Appl. 4 (1) (1996) 74–83.

[22] D. Milutinovic, V. Milutinovic, B. Soucek, The honeycomb architecture, IEEE Comput. 20 (4) (1987) 81–83.

[23] S. Stojanović, D. Bojić, V. Milutinović, Solving gross Pitaevski equation using dataflow paradigm, IPSI BgD Trans. Internet Res. 9 (2) (2013).

[24] W.-Y. Chung, J.-H. Yoo, Remote water quality monitoring in wide area, Sens. Actuators B Chem. 217 (2015) 51–57.

[25] Waterwatch, 2020. Available online: https://waterwatch.usgs.gov/wqwatch/. (accessed on February, 2020).

[26] S.K. Priya, G. Shenbagalakshmi, T. Revathi, Design of smart sensors for real time drinking water quality monitoring and contamination detection in water distributed mains, Intl. J. Eng. Technol. 7 (1) (2018) 47–51.

[27] H.A. Rahim, S.N. Zulkifli, N.A.M. Subha, R.A. Rahim, H.Z. Abidin, Water quality monitoring using wireless sensor network and smartphone-based applications: a review, Sens. Transducers 209 (2) (2017) 1–11.

[28] S. Venkatramanan, S.Y. Chung, S.Y. Lee, N. Park, Assessment of river water quality via environmentric multivariate statistical tools and water quality index: a case study of Nakdong River Basin, Korea, Carpathian J. Earth Environ. Sci. 9 (2) (2014) 125–132.

[29] R. Yue, T. Ying, A water quality monitoring system based on wireless sensor network & solar power supply, in: 2011 IEEE International Conference on Cyber Technology in Automation, Control, and Intelligent Systems, Kunming, 2011, pp. 126–129.

[30] S. Geetha, S. Gouthami, Internet of things enabled real time water quality monitoring system, Smart Water Int. J. 2 (1) (2017). Springer Open.

[31] F. Williamson, et al., Online water quality monitoring in the distribution network, Water Pract. Technol. 9 (4) (2014) 576–585. IWA Publishing.

[32] M. Saravanan, A. Das, V. Iyer, Smart water grid management using LPWAN IoT technology, in: 2017 Global Internet of Things Summit (GIoTS), Geneva, 2017, pp. 1–6.

[33] K.A.U. Menon, P. Divya, M.V. Ramesh, Wireless sensor network for river water quality monitoring in India, in: 2012 Third International Conference on Computing, Communication and Networking Technologies (ICCCNT'12), Coimbatore, 2012, pp. 1–7.

[34] C. Alippi, C. Galperti, M. Roveri, A robust, adaptive, solar-powered WSN framework for aquatic environmental monitoring, IEEE Sens. J. 11 (1) (2011) 45–55.

[35] A. Faustine, A.N. Mvuma, H.J. Mongi, M.C. Gabriel, A.J. Tenge, S.B. Kucel, Wireless sensor networks for water quality monitoring and control within lake Victoria basin: prototype development, Wirel. Sens. Netw. 6 (2014) 281–290.

[36] J. Wang, X. Ren, Y. Shen, S. Liu, A remote wireless sensor networks for water quality monitoring, in: 2010 International Conference on Innovative Computing and Communication and 2010 Asia-Pacific Conference on Information Technology and Ocean Engineering, Macao, 2010, pp. 7–12.

[37] S.B. Chandnapalli, S. Reddy, R. Lakshmi, Design and deployment of aqua monitoring system using wireless sensor networks and IAR_Kick, J. Aquac. 5 (2014) iss.7.

[38] F. Adamo, F. Attivisimo, C.G.C. Carducci, A.M.L. Lanzolla, A smart sensor network for sea water quality monitoring, IEEE Sens. J. 15 (5) (2015) 2514–2522.

[39] D.S. Simbeye, S.F. Yang, Water quality monitoring and control for aquaculture based on wireless sensor networks, J. Networks 9 (4) (2014) 840–849.

[40] G. Cario, A. Casavola, P. Gjanci, M. Lupia, C. Petrioli, D. Spaccini, Long lasting underwater wireless sensors network for water quality monitoring in fish farms, in: OCEANS 2017—Aberdeen, Aberdeen, 2017, pp. 1–6.

[41] P. Jiang, H. Xia, Z. He, Z. Wang, Design of a water environment monitoring system based on wireless sensor networks, Sensor 9 (8) (2009) 6411–6434.

[42] B. Ngom, M. Diallo, B. Gueye, N. Marilleau, LoRa-based measurement station for water quality monitoring: Case of Botanical Garden Pool, in: 2019 IEEE Sensors Applications Symposium (SAS), Sophia Antipolis, France, 2019, pp. 1–4.

[43] K.M. Simitha, M.S. Subodh Raj, IoT and WSN based water quality monitoring system, in: 2019 3rd International conference on Electronics, Communication and Aerospace Technology (ICECA), Coimbatore, India, 2019, pp. 205–210.

[44] Y. Ma, W. Ding, Design of intelligent monitoring system for aquaculture water dissolved oxygen, in: 2018 IEEE 3rd Advanced Information Technology, Electronic and Automation Control Conference (IAEAC), Chongqing, 2018, pp. 414–418.

[45] P.D. Gennaro, D. Lofu, D. Vitanio, P. Tedeschi, P. Boccadoro, WaterS: a Sigfox-compliant prototype for water monitoring, Internet Technol. Lett. 2 (1) (2019).

[46] J. Dong, G. Wang, H. Yan, J. Xu, X. Zhang, A survey of smart water quality monitoring system, Environ. Sci. Pollut. Res. 22 (7) (2015) 4893–4906. SpringerLink.

[47] A.-M. Dunca, Water pollution and water quality assessment of major transboundary rivers from Banat (Romania), Hindawi J. Chem. (2018).

[48] N.A. Cloete, R. Malekian, L. Nair, Design of smart sensors for real-time water quality monitoring, IEEE Access 4 (2016) 3975–3990.

[49] A. Francesco, A. Filippo, G.C. Carlo, M.L. Anna, A smart sensor network for sea water quality monitoring, IEEE Sens. J. 5 (5) (2015) 2514–2522.

[50] B. O'Flynn, et al., SmartCoast: a Wireless sensor network for water quality monitoring, in: 32nd IEEE Conference on Local Computer Networks (LCN 2007), Dublin, 2007, pp. 815–816.

[51] S. Oberoi, K.S. Daya, P.S. Tirumalai, Microwave sensor for detection of *E. coli* in water, in: 2012 Sixth International Conference on Sensing Technology (ICST), Kolkata, 2012, pp. 614–617.

[52] J.C. Ritchie, P.V. Zimba, J.H. Everitt, Remote sensing techniques to assess water quality, Photogram. Eng. Remote Sens. 69 (6) (2003) 695–704.

[53] Water Sanitation Health, 2020. Available online: https://www.who.int/water_sanitation_health/dwq/fulltext.pdf. (accessed on February, 2020).

[54] D. Maher, IoT for Fresh Water Monitoring, KTH Royal Institute of Technology, Stockholm, Sweden, 2018.

[55] P. Kruse, Review on water quality sensors, J. Phys. D Appl. Phys. 51 (20) (2018).

[56] J. Bhardwaj, K.K. Gupta, R. Gupta, A review of emerging trends on water quality measurement sensors, in: 2015 International Conference on Technologies for Sustainable Development (ICTSD), Mumbai, 2015, pp. 1–6.

[57] Turbidity Sensor, 2020. https://wiki.dfrobot.com/Turbidity_sensor_SKU__SEN0189. (accessed on February, 2020).

[58] Waspmote Documentation, 2020. http://www.libelium.com/development/waspmote/documentation/smart-water-board-technical-guide/. (accessed on February, 2020).

[59] C.E.H. Curiel, V.H.B. Baltazar, J.H.P. Ramirez, Wireless sensor network for water quality monitoring: prototype design, WASET Intl. J. Environ. Ecol. Eng. 10 (2) (2016) 162–167.

[60] F.C. Carlos, R. Buenrostro, A.G. Ibanez, F. Estrada, Performance evaluation of an 802.15.4 wireless sensor network on a coastal environment, Int. J. Interact. Mob. Technol. 11 (1) (2017).

[61] C.C. Kao, Y.S. Lin, G.D. Wu, C.J. Huang, A comprehensive study on the internet of underwater things: applications, challenges, and channel models, Sensors 17 (7) (2017) 1477.

[62] I.F. Akyildiz, P. Wang, S.-C. Lin, SoftWater: software-defined networking for next-generation underwater communication systems, Ad Hoc Netw. 46 (2016) 1–11.

[63] M. Zennaro, et al., On the design of a Water Quality Wireless Sensor Network (WQWSN): an application to water quality monitoring in Malawi, in: 2009 International Conference on Parallel Processing Workshops, Vienna, 2009, pp. 330–336.

[64] Ubidots, 2020. Available online: https://ubidots.com/. (accessed on February, 2020).

[65] Optiqua EventLab, 2020. Available online: http://www.optiqua.com. (accessed on February, 2020).

[66] Y. Wei, J. Heidemann, D. Estrin, Medium access control with coordinated adaptive sleeping for wireless sensor networks, IEEE/ACM Trans. Netw. 12 (3) (2004) 493–506.

[67] Wsense, 2020. Available online: https://wsense.it/. (accessed on February, 2020).

[68] A.M. Manoharan, V. Rathinasabapathy, Smart water quality monitoring and metering using Lora for Smart Villages, in: 2018 2nd International Conference on Smart Grid and Smart Cities (ICSGSC), Kuala Lumpur, 2018, pp. 57–61.

[69] G. Gualtieri, et al., An integrated low-cost traffic and air pollution monitoring platform to assess vehicles' air quality impact in urban areas, Transp. Res. Procedia. 27 (2017) 609–616.

[70] A.M. Popoola, et al., Use of networks of low cost air quality sensors to quantify air quality in urban settings, Atmos. Environ. 194 (2018) 58–70.

[71] B. Fang, Q. Xu, T. Park, M. Zhang, AirSense: an intelligent home-based sensing system for indoor air quality analytics, in: UBICOMP'16, Heidelberg, Germany, 2016, pp. 109–119.

[72] News-room Fact-sheets, 2020. https://www.who.int/news-room/fact-sheets/detail/ambient-(outdoor)-air-quality-and-health. (Accessed on February, 2020).

[73] E.G. Snyder, et al., The changing paradigm of air pollution monitoring, Environ. Sci. Technol. 47 (20) (2013) 11369–11377.

[74] A. Dobre, et al., Flow field measurements in the proximity of an urban intersection in London, UK, Atmos. Environ. 39 (2005) 4647–4657.

[75] M. Pavani, P.T. Rao, Urban air pollution monitoring using wireless sensor networks: a comprehensive review, Int. J. Commun. Netw. Inform. Sec. 9 (3) (2017) 439–449.

[76] X. Fang, I. Bate, Issues of using wireless sensor network to monitor urban air quality, in: Proceedings of the First ACM International Workshop on the Engineering of Reliable, Robust, and Secure Embedded Wireless Sensing Systems, Delft, Netherland, 2017, pp. 32–39.

[77] D. Yaswanth, S. Umar, A study on pollution monitoring system in wireless sensor networks, Int. J. Comput. Sci. Eng. Technol. 3 (2013) 324–328.

[78] P.J.A. John, Wireless Air Quality and Emission Monitoring, Master Thesis, Uppsala University, 2016.

[79] A. Anjomshoaa, B. Duarte, D. Rennings, T. Matarazzo, P. de Souza, C. Ratti, City Scanner: building and scheduling a mobile sensing platform for smart city services, IEEE IoT J. (2018) 2327–4662.

[80] S. Kaivonen, E.C.-H. Ngai, Real-time air pollution monitoring with sensors on city bus, Digit. Commun. Netw. 6 (1) (2020) 23–30.

[81] A. Lavric, V. Popa, Air quality monitoring system based on large-scale WSN: a step towards a smart city, Res. Sci. Today 1 (11) (2016) 98–107.

[82] Y. Feng, H. Junyi, A. WeiPeng, C. Flanagan, C. MacNamee, S. McGrath, API monitor based on Internet of Things technology, in: 2018 12th International Conference on Sensing Technology (ICST), Limerick, 2018, pp. 213–216.

[83] S. Johnston, et al., City scale particulate matter monitoring using LoRAWAN based air quality IoT devices, Sensors 19 (1) (2019) 209–229.

[84] K. Zheng, S. Zhao, Z. Yang, X. Xiong, W. Xiang, Design and implementation of LPWA-based air quality monitoring system, IEEE Access 4 (2016) 3238–3245.

[85] M. Benammar, A. Abdaoui, H.M.S. Ahmad, F. Touati, A. Kadri, Real-time indoor air quality monitoring through wireless sensor network, Int. J. Internet Things Web Serv. 2 (2017) 7–13.

[86] J. Lozano, J.I. Suarez, P. Arroyo, J.M. Ordiales, F. Alvarez, Wireless sensor network for indoor air quality monitoring, Chem. Eng. Trans. 30 (2012).

[87] T. Alhmiedat, G. Samara, A low cost ZigBee sensor network architecture for indoor air quality monitoring, Int. J. Comput. Sci. Inform. Secur. 15 (1) (2017).

[88] Z. Tafa, F. Ramadani, Design of a multi-hop wireless network to continuous indoor air quality monitoring, J. Commun. 14 (3) (2019).

[89] S. Bhattacharya, S. Sridevi, R. Pitchiah, Indoor air quality monitoring using wireless sensor network, in: 2012 Sixth International Conference on Sensing Technology (ICST), Kolkata, 2012, pp. 422–427.

[90] G. Marques, R. Pitafrma, An indoor monitoring system for ambient assisted living based on internet of things architecture, Int. J. Environ. Res. Public Health 13 (11) (2016).

[91] M. Benammar, A. Abdaoui, S.H.M. Ahmad, F. Touati, A. Kadri, A modular IoT platform for real-time indoor air quality monitoring, Sensors 18 (2) (2018).

[92] F. Ramos, S. Trilles, A. Munoz, J. Huerta, Promoting pollution-free routes in smart cities using air quality sensor networks, Sensors 18 (8) (2018) 2507.

[93] T.Q. Ngoc, J. Lee, K.J. Gil, K. Jeong, S.B. Lim, An ESB based micro-scale urban air quality monitoring system, in: 2010 IEEE Fifth International Conference on Networking, Architecture, and Storage, Macau, 2010, pp. 288–293.

[94] EPA Air Quality Index: A Guide to Air Quality and Your Health, 2014. Available online: https://www3.epa.gov/airnow/aqi_brochure_02_14.pdf. (accessed on February, 2020).

[95] EPA, 2020. Available online: https://www.epa.gov/. (accessed on February, 2020).

[96] P. Bierwidth, Carbon Dioxide Toxicity and Climate Change: A Serious Unapprehended Risk For Human Health, Australian National University, 2016. Available online: https://pdfs.semanticscholar.org/a53f/3a6f7a0db8cfd006c51f19c76a 68f3386f7e.pdf?_ga=2.153204347.29371009.1584524680-264660736.1584524680. (accessed on February, 2020).

[97] A. G-Azam, B. R-Zanjani, M. B-Mood, Effects of air pollution on human health and practical measures for prevention in Iran, J. Res. Med. Sci. 21 (5) (2016).

[98] Z. Yunusa, M.N. Hamidon, A. Kaiser, Z. Awang, Gas sensors: a review, Sensors Transducers 168 (4) (2014) 61–75.

[99] P. Kumar, L. Morawska, C. Martani, G. Biskos, M. Neophytou, S.D. Sabatino, M. Bell, L. Norford, R. Britter, The rise of low-cost sensing for managing air pollution in cities, Environ. Int. 75 (2015) 199–205.

[100] Libelium, 2018. Available online: http://www.libelium.com/downloads/ documentation/gases_sensor_board_pro.pdf. (Accessed on January, 2018).

[101] HEI Review Panel on Ultrafine Particles, Understanding the Health Effects of Ambient Ultrafine Particles, Health Effects Institute, Boston, MA, 2013. Available online: https://www.healtheffects.org/system/files/Perspectives3.pdf. (accessed on February, 2018).

[102] S. Abraham, X. Li, A cost-effective wireless sensor network system for indoor air quality monitoring applications, Procedia Comput. Sci. 34 (2014) 165–171.

[103] V. Hejlova, V. Voženilek, Wireless sensor network components for air pollution monitoring in the urban environment: criteria and analysis for their selection, Wirel. Sens. Netw. 5 (2013) 229–240.

[104] W.Y. Yi, K.M. Lo, T. Mak, K.S. Leung, Y. Leung, M.L. Meng, A survey of wireless sensor network based air pollution monitoring systems, Sensors 15 (12) (2015) 31392–31427.

[105] A.T. Campbell, S.B. Eisenman, N.D. Lane, E. Miluzzo, R.A. Peterson, H. Lu, X. Zheng, M. Musolesi, K. Fodor, G.S. Ahn, The rise of people-centric sensing, IEEE Internet Comput. 12 (4) (2008) 12–21.

[106] A. Kapadia, D. Kotz, N. Triandopoulos, Opportunistic sensing: security challenges for the new paradigm, in: 2009 First International Communication Systems and Networks and Workshops, Bangalore, 2009, pp. 1–10.

[107] A.-S. Mihaita, L. Dupont, O. Cherry, M. Camargo, C. Cai, Air quality monitoring using stationary versus mobile sensing units: a case study from Lorraine, France, in: 25th ITS World Congress 2018, Sep 2018, Copenhagen, Netherlands, 2018, pp. 1–11.

[108] Airqualityegg, 2020. Available online: https://airqualityegg.com/home. (Accessed on February, 2020).

[109] M. Nyarku, M. Mazaheri, R. Jayaratne, M. Dunbabin, M.M. Rahman, E. Uhde, L. Morawska, Mobile phones as monitors of personal exposure to air pollution: is this the future? PLoS One 13 (2) (2018) 1–18.

[110] B. Hull, et al., A distributed mobile sensor computing system, in: Proceedings of the 4th International Conference on Embedded Networked Sensor Systems, ACM, Boulder, CO, 2006, pp. 125–138.

[111] S. Mathur, T. Jin, N. Kasturirangan, J. Chandrasekaran, W. Xue, M. Gruteser, W. Trappe, Parknet: drive-by sensing of road-side parking statistics, in: Proceedings of the 8th International Conference on Mobile Systems, Applications, and Services, ACM, San Francisco, CA, 2010, pp. 123–136.

[112] M. Wang, R. Birken, S.S. Shamsabadi, Framework and implementation of a continuous network-wide health monitoring system for roadways, in: Nondestructive Characterization for Composite Materials, Aerospace Engineering, Civil Infrastructure, and Homeland Security 2014, vol. 9063, 2014.

[113] L.A. Akinyemi, T. Maknjoula, O. Shoqwu, C.O. Foloruns, Smart city and vehicle pollution monitoring, using wireless network system, Urban Des. 1 (2018) 48–54.

[114] Perkin Elmer Sensor, 2020. Available online: http://www.aqmd.gov/aq-spec/product/perkin-elmer---elm. (accessed on February, 2020).

[115] Labview, 2018. Available online: http://www.ni.com/en-rs/shop/labview.html. (accessed on February, 2018).

[116] Linksprite, 2018. Available online: http://www.linksprite.com/linksprite-pcduino3. (accessed on February, 2018).

[117] Calvert Ventures LLC, TerraView—A Platform to View, Analyze and Control Your IoT World, Calvert Ventures LLC, California, USA, 2015.

[118] M.M. Maricq, D.H. Podsiadlik, R.E. Chase, Examination of the size-resolved and transient nature of motor vehicle particle emissions, Environ. Sci. Technol. 33 (10) (1999) 1618–1626.

[119] Emoncms, 2020. Available online: https://emoncms.org. (accessed on February, 2020).

[120] Technical Assistance Document for the Reporting of Daily Air Quality—the Air Quality Index (AQI) by U.S. Environmental Protection Agency, 2018. Available online: https://www3.epa.gov/airnow/aqi-technical-assistance-document-sept2018.pdf. (accessed on February, 2020).

[121] World Health Organization, WHO Guidelines for Indoor Air Quality: Selected Pollutants, World Health Organization, Geneva, Switzerland, 2010.

[122] D. Burke, M. Estrin, A. Hansen, N. Parker, S. Ramanthan, V.S. Reddy, M.B. Srivastava, Participatory sensing, in: SenSys'06, 2006, Boulder, Colorado, USA, 2006.

[123] Pycom, 2020. Available online: https://pycom.io/product/sipy/. (accessed on February, 2020).

[124] V. Blagojevic, et al., A systematic approach to generation of new ideas for PhD research in computing, Adv. Comput. 104 (2016) 1–19.

[125] R. Williams, et al., Sensor Evaluation Report, US Environmental Protection Agency, Washington, DC, 2014.
[126] I. Tsekoura, G. Rebel, P. Glösekötter, M. Berekovic, An evaluation of energy efficient microcontrollers, in: 2014 9th International Symposium on Reconfigurable and Communication-Centric Systems-on-Chip (ReCoSoC), Montpellier, 2014, pp. 1–5.
[127] Y. Han, B. Park, J. Jeong, A novel architecture of air pollution measurement platforms using 5G Blockchain for industrial IoT applications, Procedia Comput. Sci. 155 (2019) 728–733.
[128] G. Hagler, M. Freeman, Real-Time Geospatial (RETIGO) data viewer: a web-based tool for data exploration, in: EPA Air Sensors Workshop, Durham, NC, 2015.

About the author

Zhilbert Tafa is currently a Professor with the School of Computer Science and Engineering, University for Business and Technology in Kosovo. He obtained his PhD in Computer Engineering and Information Theory from Belgrade University, Serbia; and both BS and the MS degrees in Electrical and Computer Engineering from the University of Montenegro. Prof. Tafa has been a collaborator on several APEG (Applied Electronics Group) projects, at the University of Montenegro.

As a member of a number of renowned scientific journal and conference boards, over the past years Dr. Tafa has also been contributing as a reviewer of scientific papers, including a number of IEEE magazines and conferences. He received the most cited paper award from IEEE MECO conference, 2017. His current research interests involve areas of networking protocols, wireless sensor networks, IoT infrastructure, and machine learning.

CHAPTER FOUR

Energy efficient implementation of tensor operations using dataflow paradigm for machine learning [☆]

Miloš Kotlar[a], Marija Punt[a], and Veljko Milutinović[b]
[a]School of Electrical Engineering, University of Belgrade, Belgrade, Serbia
[b]Department of Computer Science, Indiana University in Bloomington, IN, United States

Contents

[☆]This article presents a follow-up research of an article published in International Supercomputing conference workshop proceedings, ExaComm 2018, Frankfurt.

Advances in Computers, Volume 126
ISSN 0065-2458
https://doi.org/10.1016/bs.adcom.2021.11.011

Abstract

This article introduces 10 different tensor operations, as well as their generalizations and implementations within the dataflow paradigm. Tensor operations can be utilized for addressing a number of big data problems in machine learning and data mining, such as deep learning, clustering, classification, dimension reduction, anomaly detection, and applications in civil and geo-engineering. With proliferation of data and devices, the main challenge is finding a way to process big quantities of data, especially in environments where resources are limited. This article sheds light on the energy efficient dataflow implementation of tensor operations used in machine learning. The iterative nature of tensor operations and a large amount of data makes them suitable for the dataflow paradigm. All dataflow implementations are analyzed comparatively with the related control-flow implementations, for speedup, complexity, power savings, and meantime between failures. The core contribution of this article is to classify existing implementations of tensor operations on four main architectural approaches in environments with limited resources and to propose energy efficient implementation of 10 tensor operations used in machine learning algorithms. The proposed implementations on the dataflow paradigm are compared against the traditional control-flow paradigm for various data set sizes, and in various conditions of interest.

1. Introduction

The volume of data has exponentially increased over the last few years: from 4 zettabytes in 2013 to 44 zettabytes today [1]. Based on predictions [2] the velocity of data will experience exponential growth, and by 2025 the volume of big data will be 165 zettabytes. With such exponential growth of data volume and velocity, the big data are becoming the essential asset across the entire spectrum of different domains, from healthcare to smart farming [3,4]. Level of data organization and structure mostly depends on a particular domain, use case, and data variety. The biggest problem is how to obtain value from the data [5].

Data mining techniques used for finding hidden knowledge in the data rely on machine learning algorithms. Machine algorithms are based on tensor operations that can be compute-intensive, and depending on the data volume they often require a large amount of resources [6] in terms of processing time and electrical power. Architectural approaches like multicore processors (central processing unit) and manycore processors (graphics processing unit) can process enormous amount of data, and thus present a ground base for machine learning algorithms. Most of the existing solutions are based on anticipated cloud-based architectures, which are expected to be hybrid cloud-edge architectures suitable for IoT and WSN [7]. Cloud-based architectural approaches rely on multicore and manycore processors with huge processing power, while edge-based architectural approaches rely on field programmable gate arrays and application-specific integrated circuits, which makes it possible to process data without latency and with low power dissipation. The expansion of machine learning algorithms [8,9,10] and the rapid development of the Internet of Things (IoT) and Wireless Sensor Networks (WSN) [11] have created a strong need for alternative architectural approaches.

The origin of the tensor calculus dates back to the beginning of the last century. The first well-known concept of the tensor was introduced in Einstein's theory of relativity [12]. After that, tensors have been used in many fields such as physics, quantum mechanics, quantum chemistry, engineering, and only more recently in machine learning. Nowadays, many software packages are based on multilinear algebra, which brings tensor operations into focus as an approach for efficient data encoding that offers optimizations for orders of magnitude [13]. This article analyzes applications of tensor operations in machine learning algorithms and introduces low power implementations using the dataflow paradigm. Tensors can be used in various fields of artificial intelligence such as natural language processing, for encoding and processing symbol structures in distributed neural networks [14], and in deep neural networks, for describing parameters and relations between neurons in recurrent networks [15]. When Google launched the TensorFlow library for machine learning based on tensors that enabled executions on tensor processing units, artificial intelligence witnessed a meteoric rise [16,17,18].

The dataflow paradigm relies on the Feynman paradigm in which an algorithm can be converted into an execution graph where each node in the graph presents one arithmetic or logic unit. Each arithmetic or logic unit can compute only one operation at a time. Such an approach perfectly fits the field-programmable gate array (FPGA) architecture which enables low power execution [19]. The dataflow paradigm can be referred to as computing in space where arithmetic and logic units are placed dimensionally on an FPGA card. The first steps towards mapping of algorithms onto hardware can be found in Ref. [20], while the details of a current modern approach can be found in Ref. [21]. The conventional control-flow paradigm can be referred to as computing in time, where operations are computed at different moments in time in the same arithmetic and logic units [22]. The speedup achieved using the dataflow paradigm depends on the characteristics of loops and the amount of time that the algorithm spends executing the loops, while power savings depend on the clock frequency and execution graph complexity.

According to the open literature there is a lack of articles related to energy efficient implementations of tensor operations for machine learning in environments with limited resources and thus this article presents 10 basic tensor operations for machine learning algorithms on the dataflow architecture and introduces their energy efficient implementation. Compared to conventional control–flow implementations, the proposed dataflow implementations can provide superior energy efficiency acceleration, achieving speedup per watt. These implementations can be used where computations need to be done in edge environments where power resources are limited, for example in embedded systems. This follow–up article focuses on applications of tensor operations in machine learning algorithms. It re-evaluates obtained results and provides details of optimization constructs.

Section 2 discusses the main issues of the computational aspect of machine learning algorithms, including how and why the importance of the problem will grow over time. Section 3 presents 10 basic tensor operations and discusses where it is used and which optimization constructs exist in open literature. Section 4 highlights the key advantages of the dataflow paradigm and compares the dataflow paradigm against the control–flow paradigm across different criteria such as speed, power dissipation, complexity, and meantime between failures. Section 5 introduces energy efficient implementations of 10 tensor operations and discusses optimization constructs for the dataflow paradigm. Section 6 evaluates the obtained performance and compares it with control–flow implementation. Section 7 concludes the article by discussing the achievements and benefits of the study, and ends with outlining potential topics for further research.

2. Compute-intensive machine learning algorithms

With the rapid expansion of data and devices, machine learning algorithms have become widely used for prediction of stock prices [23], monitoring patients' health condition [24], financial fraud detection [25], anomaly detection [26], and other domains [27,28,29]. Machine learning algorithms such as neural networks consist of compute-intensive training and often strict latency inference. This section discusses different environments where training and inference could be done and how they are related to four main architectural approaches.

2.1 Cloud computing and strict latency demands

Compute-intensive training in the cloud environment is often done with unlimited power resources [30]. In such an environment resources can be scaled in order to fit massive workload demands. In environments with limited resources such as embedded systems, it is hard to perform compute-intensive training due to lack of computational power, but inference with pre-trained model can be performed. In embedded systems or real-time applications with strict latency demands, it is impossible to train a model in a cloud environment, since the processing can be done hundreds of kilometers away [31]. Also, in such environments, power and computational resources are often insufficient to perform compute-intensive training.

In order to maximize the performance of machine learning algorithms in terms of execution speed and power consumption, in environments with limited computational and power resources, this section discusses different architectural approaches for various deployment locations. The following section discusses algorithmic optimizations of tensor operations that are essential for machine learning algorithms.

2.2 Overview of different architectural approaches

In the last years, the ratio of data volume increase has been higher than the ratio of processing power increase. With the expansion of data-collecting technologies like WSN and IoT, the data volume growth ratio is also expected to continue to increase. According to Moore's law, conventional microprocessor technology based on the control-flow paradigm has been increasing the clock rate, and the processing power and power consumption have thus also increased [32]. By increasing the clock rate of high-performance systems, power dissipation also increases; consequently, the performance in terms of power dissipation of the system with limited resources decreases. In order to improve computational power, existing multicore microprocessors exploit the advantages of control-flow optimization constructs such as several levels of instructions and data caches, memory and I/O management, and branch and data predictors. These optimization constructs increase the number of transistors on a chip, thereby increasing power dissipation.

The fact that technology limits are almost reached [33] led to the development of alternative architectural approaches. One of such approaches involves adding more processors to increase processing power. That approach renders better performance than a single processor solution, but requires algorithmic changes in order to split processing over more units.

In contrast to multicore microprocessors known as CPU, an alternative approach for fast computing is manycore microprocessing known as GPU. The GPU approach is better for big data streaming algorithms, such as machine learning algorithms which rely on tensor operations. Tensor operations can exploit architectural constructs such as fast shared memories, which results in a much better performance compared to manycore microprocessors. For example, the results presented in article [34] show that the GPU implementations of sorting algorithms with the use of shared memory are two times faster than implementations that use only global memory of a device, and up to 7.5 times faster than CPU implementation. While GPUs operate at lower frequencies, they typically have many times the number of cores compared with conventional CPUs. GPUs can therefore process far more data per second. Nowdays GPU stands for state-of-the-art high performance computing systems. The main disadvantage of such an approach is large power dissipation, which is crucial in environments with limited resources. Owing to a large number of transistors, power dissipation can be enormously high, especially for machine learning algorithms that operate with big data.

In contrast to controlflow approaches, compilation in the dataflow approach goes to a much lower hardware level. In addition to energy efficient acceleration, dataflows are powerful because they are adaptable and make it easy to implement changes by reusing an existing chip. Most emerging technologies require an increase in processing power capabilities [35,36]. There are a number of dataflow implementations of machine learning algorithms in open literature. Huang et al. [37] show that FPGA outperforms GPU when using pruned or compact data types versus full 32 bit floating point data for deep neural networks. When using the FPGA architecture, neural networks run faster than on GPUs on small (32×32) inputs, while consuming up to $20 \times$ less energy and power [38]. Voss et al. [39] discuss the high performance implementation of convolutional neural networks and compare the obtained performance against other implementations, showing that the proposed design reaches 2450 GOPS when running VGG16 as a test case.

In the last decade, ASIC chips have been used mostly for specific applications, such as cryptocurrency mining, where enormous computing power is needed [40]. The ASIC is customized for a particular use, rather than general-purpose use [41]. Using the ASIC specially designed for an algorithm definitely outperforms most of the above-mentioned paradigms under conditions of interest. Two main advantages of the ASIC chips are: (A) better speedup, and (B) less electric power that is required in order to operate. Despite these advantages, the ASIC chip does not have the flexibility to run a number of algorithms. The ASIC is hard to design; moreover, it is very expensive and time consuming if it needs to be redesigned. For example, it may take up to 18 months to manufacture one ASIC chip [42].

The development of high-performance computing systems is still based on the control-flow paradigm with powerful multicore and manycore processors, which offers the best solution for training and inference of machine learning algorithms in environments where resources are unlimited and where execution time is the primary concern. Currently, machine learning based applications use GPUs for the cloud environment, whereas FPGA, ASIC, and embedded processors of any type are on the rise in the edge environment. Cloud high-performance computing offers immense opportunities but also struggles with many issues which still have to be addressed [43]. Emerging low precision computations, irregular parallelism, custom data types, and energy efficient computations are now shifting data and processing significantly away from the cloud to end devices [7,37]. In such environments, alternative architectural principles such as the dataflow paradigm can achieve significantly better performance compared to the control-flow paradigm when energy efficient computation is the prime concern.

3. Basic tensor operations for machine learning

This section presents tensor operations and how they are used machine learning algorithms and discusses existing implementations according to the open literature. After that, a classification table is given which shows a lack of research articles with energy efficient dataflow implementations.

3.1 Introduction to tensor operations

In mathematics, tensor is an object that describes the relation between sets of objects related to vector space. In physics, tensor is an object that operates on a vector to produce another vector [44]. In computer science, tensor is an n dimensional array that represents data. The rank or order of a tensor is determined by the number of axes. Tensors can take several different forms depending on order, for example: scalar is a zero-order tensor, vector is a first-order tensor, matrix is a second-order tensor, cube is a third-order tensor.

Machine learning relies on linear algebra and tensor operations [45]. Almost every machine learning algorithm computes tensor operations in order to produce the result. Tensors are good for high-dimensional data presentation, especially when referring to neural network data representation. Aside from storing data, tensors also include data relations and can perform linear transformations. From a computer science perspective, this is the reason why they can be considered an object instead of a simple data structure.

In order to better understand tensors and their operations, examples are presented for orders from zero to four. A scalar is a zero-dimensional tensor with a single number and the rank of zero. A vector is a single-dimension tensor often referred to as an array with the rank of one. A commonly used tensor in machine learning algorithms is the rank of two, referred to as a matrix; for example, a matrix can be a suitable structure for image representation where each entry presents one pixel. A video can be represented as a three-dimensional tensor where the first two dimensions represent a single frame in a video, while the third dimension represents frames in a video as array of images. A facial image database is a sixth-order tensor where two dimensions of an image and the remaining four dimensions are used for describing different facial expressions related to the image [46].

The relation between tensor objects and machine learning algorithms is apparent in open literature. Neural networks use tensor objects for describing relations between neurons in a network and tensor multiplication for training and inference [15]. Tensor rank decomposition and tensor factorization methods are used from alternating optimization to stochastic gradient, in statistical performance analysis and applications ranging from mixture and topic modeling, classification, and multilinear subspace learning [47]. Tensors and their decompositions are also widely used in unsupervised learning settings, as well as other subdisciplines like temporal and multi-relational data analysis [48]. Over the last decade, tensor decompositions have been efficiently used for estimating parameters of latent variable models [49] and independent component analysis [50]. Based on an overview of open literature, this section divides basic tensor operations used in machine learning algorithms into three main categories and presents its applications, implementations, and algorithmic optimizations in terms of time complexity and memory utilization. Basic tensor operations are divided into the following groups based on their characteristics and applications in machine learning algorithms: (A) arithmetic operations, (B) transformation operations, and (C) factorization operations. High–order tensor operations are NP–hard problems [51] and this article focuses primary on second–order tensor operations which are widely used in machine learning problems with possible extension to high–order tensors.

3.2 Arithmetic tensor operations

Arithmetic operations are basic linear algebra arithmetic operations, such as tensor addition, tensor composition, tensor transpose, and divergence of a tensor field. These operations are used as general purpose operations in machine learning algorithms. For example, in neural networks, tensor addition and compositions are used for calculating results in the feed-forward phase, where inputs and weights are encoded in tensors. The tensor transpose operation is often used as a basic operation for complex tensor decompositions.

3.2.1 Tensor addition

Tensor addition is a basic arithmetic general purpose operation with time complexity of $O(n)$. In open literature there are not many improvements that accelerate the operation due to efficient computation and low implementation complexity. However, if a tensor is sparse with a number of elements equal to 0, it is efficient to keep only valuable data. Instead of keeping full data in a tensor, a map function can be used for keeping only valuable data, depending on the position [52]. By enabling new fetching and writing mechanisms for sparse tensors, memory utilization can be improved by storing only valuable data; however, such an approach introduces maintaining complexity, especially for high–order tensors. Beside algorithmic optimizations for sparse tensors, article [53] proposes new hardware mechanisms for handling sparsity into hardware to enable better bandwidth utilization and compute throughput.

3.2.2 Tensor transpose

Tensor transpose is a basic tensor operation which flips a tensor over its diagonals with time complexity of $O(n)$. This is a basic tensor operation mostly used in complex tensor decompositions where by transposing a tensor one can get an orthogonal tensor. The orthogonal tensor may also be used as a data structure for representation and manipulating graphs. In open literature there are several implementations that differ in terms of memory utilization and time complexity [54]. There is a trade-off between time complexity and memory utilization, where time complexity decreases with increased memory utilization. These implementations avoid explicitly transposing a tensor in memory by simply accessing the same data in a different order [55]. However, in some cases it is necessary to physically reorder elements in a tensor, which could allocate memory resources for high-order tensors. In such cases, performance of the algorithm can be improved by increasing memory locality in cache blocks. For example, with a tensor stored in row–major order, the rows of the tensor are contiguous in memory, while columns are discontiguous.

3.2.3 Tensor product

A tensor product is a basic arithmetic general purpose operation with high computational complexity. It is used as a basic operation in various machine learning algorithms. Applications range from training and inference neural networks to computing distances in density-based classification and clustering methods [56]. The naive implementation, also known as the schoolbook algorithm, has time complexity of $O(n^3)$. The schoolbook algorithm consists of three nested loops that iterate rows and columns. The performance of such implementation is inefficient and according to literature there are a number of papers that proposes arithmetic optimizations. For example, if this algorithm relies on multicore microprocessors and if inner loops swap their positions, the final result will be the same, but performance can be drastically improved due to aligned cache memory access, where columns or rows can be aligned with cache blocks, enabling fast access to these values. Article [57] discusses applying different caching strategies to matrix multiplication and its impact on performance.

An alternative approach to the schoolbook algorithm is the divide and conquer method, known as the Strassen algorithm [58]. Practical implementations of the Strassen algorithm switch to standard methods of matrix multiplication for small enough submatrices, for which those algorithms are more efficient. The particular crossover point for which the Strassen algorithm is more efficient depends on the specific implementation and hardware. The idea behind the Strassen algorithm is to interpret tensor multiplication as a re-cursive problem which is favorable for high-order tensors. The algorithm splits the tensor into smaller chunks and recursively computes the result. It has been observed that this crossover point has been increasing in recent years, so even a single step of the algorithm is often not beneficial in current architectures compared to a highly optimized multiplication, until tensor sizes exceed 1000 or more. A more recent study [59] observed the benefits for low-order tensors. The Strassen algorithm has time complexity of $O(n^{2.807})$, which is faster than schoolbook implementation.

The Coppersmith-Winograd algorithm [60] is a tensor multiplication algorithm with time complexity of $O(n^{2.376})$, which is better time complexity than the Strassen algorithm. Despite its low time complexity, it is rarely used in industry due to large constant factors, which makes it impractical for application. However, there are articles [61] in open literature which present algorithmic optimizations which are able to achieve even better performance.

3.2.4 Divergence, curl, and gradient

Divergence, curl, and gradient are operations commonly used in continuum mechanics and physics for transforming tensor fields into other forms of tensor fields using partial derivatives. For example, divergence [62] is a tensor operator that produces tensor field giving the quantity of a tensor field's source at each point. Gradient descent is an important optimization machine learning algorithm which finds the values of parameters of a differentiable function that minimizes the cost function. In machine learning algorithms, the goal is to estimate the function that maps input data into output results. Optimization functions are used for finding optimal coefficients for the algorithm that gives the smallest error. Machine learning algorithms such as linear regression and logistic regression have coefficients that characterize the algorithm's estimate for the estimate function. The cost function calculates the aggregated error between predictions and the actual output values. A derivative can be calculated from the cost function and coefficients so the coefficients can be updated in order to minimize the error. The stochastic gradient descent algorithm is an extension of the gradient descent algorithm which is efficient for high-order tensors [63]. From a computational perspective, divergence, curl, gradient, and gradient descent methods can be interpreted as tensor multiplication with time complexity of $O(n^3)$.

3.3 Transformation tensor operations

Transformation operations are basic linear algebra transformation operations, such as the tensor inversion process or principal and primary invariants. These operations are used in linear algebra as general purpose operations. While the tensor inversion process is widely used for estimating ordinary least squares in linear regression in machine learning algorithms, primary and principal invariants do not have direct application in machine learning algorithms but are an important concept in linear algebra.

3.3.1 Tensor inverse

Tensor inversion is a process that finds another tensor when multiplied with tensor results in an identity tensor. In computer graphics, the tensor inversion process is used in graphics rendering and simulations. In mathematics, tensor inversion can be used for solving a set of linear equations. A need for solving a set of linear equations simultaneously arises in various fields of applied science. It is used in physics and engineering for simulating fluid flows while in finance it is employed for estimating models used for certain types of option pricing [64,65]. In machine learning it is the crucial method for solving a set of linear equations mostly used in linear regression for estimating ordinary least squares [66].

Gaussian elimination [67] is an algorithm that can be used for solving systems of linear equations and calculating tensor inverse with time complexity of $O(n^3)$. The algorithm solves the problem in two phases: (A) reducing a given system to a row echelon form and (B) calculating whether there are no solutions, a unique solution, or infinitely many solutions. The second phase continues to use row operations until the solution is found. Similar to this algorithm, LU decomposition [68] interprets tensor as a product of a lower triangular tensor and an upper triangular tensor. The result of LU decomposition can be presented as the result of the Gaussian elimination algorithm.

Another algorithm that can be used for computing the inversion of a tensor is Newton's method [69]. Newton's method may be useful when an application deals with partly related tensors that behave in a similar way. The goal of the algorithm is to approximate the inverse of a tensor using the inverse of the previous tensor. This approach can be applied only to tensors which are similar to time complexity of $O(\log(n))$.

Another method for finding tensor inverse is the Cayley-Hamilton method which allows the inverse of a tensor to be expressed in terms of determinant, traces, and powers [70]. The Cayley-Hamilton method has time complexity of $O(n^3)$ and this method is rarely used due to its high implementation complexity. Blockwise inversion [71] is a divide and conquer algorithm that uses blockwise inversion to invert a tensor. This algorithm is based on tensor multiplication and its time complexity depends on the algorithm used for multiplication.

3.3.2 Primary invariants

Primary invariants in linear algebra represent coefficients of the characteristic polynomial of a tensor which do not change with the rotation of the coordinate system. In physics, primary invariants are commonly used in formulating closed-form expressions for strain energy density, also known as Helmholtz free energy [72]. Primary and principal invariants do not have direct application in machine learning algorithms, but this section nevertheless discusses their implementations since they are an important linear algebra concept. Tensor decomposition, such as LU and QR decompositions, can be efficiently used for finding primary invariants of a tensor with time complexity of $O(n^3)$. There are several methods for computing QR decomposition, such as the Gram–Schmidt process, Householder transformations, and Givens rotations, each of which has a number of advantages and disadvantages [73].

3.3.3 Principal invariants
Principal invariants are similar to primary invariants where the first principal invariant, known as trace, is always the sum of diagonal elements. The determinant of a tensor is a special case of principal invariants which encodes certain properties of the linear transformation of a tensor. Principal invariants can be used for finding eigenvalues and eigenvectors, which is discussed in the next subsection. They are often calculated using decomposition techniques due to better time complexity, especially in architectural approaches suitable for parallel computing. The most popular decomposition methods for finding principal invariants are LU decomposition, QR decomposition, and Cholesky decompositions [74].

3.4 Factorization tensor operations
Factorization operations are complex linear algebra operations such as finding eigen decomposition with eigenvalues and eigenvectors or finding the rank of a tensor. These operations are used as general purpose operations in linear algebra. All operations have an important role in machine learning algorithms. Eigen decomposition is used in principal component analysis for dimensionality reduction and anomaly detection, while finding the rank of a tensor is important in cases when it is necessary to find a tensor of a certain rank which approximates a given tensor.

3.4.1 Eigenvalues and eigenvectors
In linear algebra, eigenvalues and eigenvectors of a linear transformation is a tensor that does not change direction when a linear transformation is applied to it. In machine learning, eigenvalues and eigenvectors play an important role in methods such as Principal Component Analysis (PCA) [75]. PCA is an unsupervised statistical method used for dimensionality reduction, which is extremely important in high-dimensional data. PCA

transforms high-dimensional data into low-dimensional subspace while maintaining its characteristics and relations between them. This method can also be used for detecting anomalies in data by transforming data into low-dimensional subspace and calculating reconstruction error. If reconstruction error differs from error distribution, this instance can be considered as anomaly.

From the implementation perspective, one of the most important problems is designing an efficient and stable algorithm for finding the eigenvalues and eigenvectors of a tensor. Most algorithms in open literature are iterative: they compute eigenvalues and eigenvectors by iteratively calculating sequences that at the end converge to eigenvectors and eigenvalues. A commonly used method for finding eigenvalues and eigenvectors is QR decomposition, which has already been discussed in this article. The Jacobi algorithm [76] is another iterative method that can be used for calculating eigenvalues and eigenvectors with time complexity of $O(n^3)$. Another approach for eigen analysis with time complexity of $O((4/3)n^3 + O(n^2))$ is the divide and conquer method which divides the problem into two parts each of which is solved recursively, so the eigenvalues of the original problem are computed from the results of these smaller problems [77].

3.4.2 Spectral decomposition

Spectral decomposition, also known as eigen decomposition, is the factorization of a tensor into a canonic form in terms of eigenvalues and eigenvectors. As mentioned above, factorization components of eigen decompositions are used in various machine learning algorithms, for detecting anomalies in data and data reduction. Singular value decomposition is an algorithm used for calculating eigen decomposition with time complexity of $O(n^3)$. Singular value decomposition can be interpreted as a composition of three geometrical transformations: rotation, scaling, and another rotation. This method factorizes a tensor in its factorization components, which are the unitary tensor, the rectangular diagonal tensor with non-negative real numbers on the diagonal, and the real or complex unitary tensor. This algorithm has time complexity of $O(n^2)$. This method is also used for data compression in various fields, such as compressing health monitoring data [78], which achieves better storage efficiency.

3.4.3 Rank of a tensor

The rank of a tensor is the dimension of the subspace generated by the columns of that tensor. In machine learning, a standard problem is to find a tensor of a certain rank which approximates a given tensor and this is why finding the rank of a tensor is elaborated in this article [79]. A common approach to finding the rank of a tensor is reducing it to a simpler form, generally row echelon form, by using elementary row operations [80]. Once in the row echelon form, the rank is the same for both row rank and column rank, and equals the number of pivots and the number of non-zero rows. Algorithms such as Gaussian elimination, singular value decomposition, or QR decomposition can be used for transforming a tensor in the echelon form, which has been discussed in the previous subsections. The basic Gaussian elimination can be unreliable due to floating point computations so other methods such as singular value decomposition or QR decompositions can be used instead.

3.5 Evaluation of existing solutions

Most of the tensor operations presented in this section are compute-intensive operations which require a large amount of computational and power resources to complete a given task. Table 1 gives an overview of 10 basic tensor operations used in machine learning algorithms with

Table 1 List of 10 basic tensor operations with their theoretical time complexities of interest for machine learning algorithms, divided in three main groups.

Tensor operation	Group	Theoretical time complexity	Optimized time complexity
Tensor addition	*Arithmetic*	$O(n)$	$O(n)$
Tensor product	*Arithmetic*	$O(n^3)$	$O(n^{2.736})$
Tensor transpose	*Arithmetic*	$O(n)$	$O(n)$
Divergence of a tensor field	*Arithmetic*	$O(n^3)$	$O(n^{2.736})$
Tensor inverse	*Transformation*	$O(n^3)$	$O(log(n))$
Primary invariants	*Transformation*	$O(n^3)$	$O(n^3)$
Principal invariants	*Transformation*	$O(n^3)$	$O(n^3)$
Eigenvalues and eigenvectors	*Factorization*	$O(n^3)$	$O((4/3)n^3 + O(n^2))$
Spectral decomposition	*Factorization*	$O(n^3)$	$O(n^3)$
Rank of a tensor	*Factorization*	$O(n^3)$	$O(n^3)$

Time complexities presented into this table can be optimized by algorithmic improvements.

Table 2 List of 10 basic tensor operations of interest in machine learning and their implementations in different architectures.

Operations	Control-flow	Dataflow
Tensor addition	[52]	[81]
Tensor transpose	[54,82]	?
Tensor product	[57]	[83,84]
Divergence, curl, and gradient	[63]	?
Tensor inverse	[67,68]	[85,86]
Primary invariants	[72]	?
Principal invariants	[74]	?
Eigenvalues and eigenvectors	[77]	[87,88]
Spectral decomposition	[78]	?
Rank of a tensor	[80]	?

theoretical time complexities for widely used implementations as well as optimized implementations with their time complexities.

There are a number of articles in open literature that discuss the implementation of tensor operations using different architectural approaches. Table 2 presents an overview and classification of basic 10 tensor operations and their implementations on different architectures. However, for some tensor operations there are no articles in open literature which discuss energy efficient implementations of tensor operations.

The following section presents the dataflow paradigm, its implementation details, and key advantages when compared against the control-flow paradigm. The article then discusses existing energy efficient implementations of listed tensor operations from open literature and proposes energy efficient dataflow-based implementations with performance evaluations and optimization constructs.

4. Dataflow paradigm

This section introduces the basic concepts of the dataflow paradigm, presents their key advantages and compares them against the conventional control-flow paradigm. It also discusses why such paradigm had been chosen for energy efficient implementations of tensor operations of interest for machine learning algorithms.

4.1 Essence of the dataflow paradigm

In essence, the dataflow card used in this article serves as an FPGA-based accelerator [22]. The dataflow card is connected to a host machine via an interconnect PCI express bus which is used for sending data from the host machine to the accelerator and retrieving the results back to the host machine. Most of the execution time in tensor operations is spent in loops. As a loop migrates from the host machine to the dataflow card, the data has to be moved to the accelerator's memory using data streams. It is possible to transfer multiple streams of data simultaneously to one dataflow card as well as to connect multiple dataflow cards via high-bandwidth links which enable a high level of parallelism.

The dataflow compiler uses kernel and manager files in order to generate an execution graph. The kernel file describes the graph structure in terms of arithmetic and logic units, including pipelines between them. The manager file controls the movement of data between the host machine and the dataflow card, as well as data movements between arithmetic and logic units in the execution graph. The language used for writing kernel and manager files is MaxJ, which is a superset of the standard Java language, extended with new functionalities for the dataflow paradigm. The main difference between the standard Java programming language and MaxJ programming language is in variables. MaxJ contains two types of variables: (A) software variables which are standard Java variables used during compile time, and (B) hardware variables which exist during the execution process and flow between arithmetic and logic units in the execution graph. Using such high level programming language for low level programming enables one to focus on solving a specific problem by leaving all low level management tasks for the compiler. This is one of the main reasons why such an approach is used in this article for evaluation tensor operations on FPGA-based accelerators.

4.2 Compilation process of the dataflow program

The host program compiles program down to the machine code level and executes on a machine based on the von Neumann architecture. The host program utilizes an accelerator for big data computations by sending data to and receiving results from the accelerator. The host program can be written in any language, including C, C++, Python, and R, and also in different environments and platforms such as MathLab, Hadoop, and MapReduce.

The dataflow compiler, referred to as MaxCompiler, first creates a graph based on the kernel files which consist of pipelines with arithmetic and logic units. Using third-party tools provided by FPGA card vendors,

MaxCompiler then converts the graph into a .max file. Once the .*max* file is created, the file is linked with the host code which represents an executable file.

At the beginning of the execution process, the manager checks if the FPGA card is configured according to the .*max* file. If it is not, the dataflow card starts the configuration process where the goal is to map the execution graph onto the FPGA card. Once the configuration process is completed, the data stream starts and the execution process begins. After that, each time when the program runs again, there is no need for reconfiguring the FPGA card. If the same FPGA card is used for another algorithm, it is necessary to reconfigure the card to match the new execution graph of an algorithm, except when both applications use a different part of the card [89].

4.3 Key differences between control-flow and dataflow paradigms

In order to present and compare the key differences between control-flow and dataflow paradigms, this article simplifies the paradigms and compares them as described below. In the conventional control-flow paradigm, the source code is transformed into a list of instructions which is then loaded into memory. The arithmetic and logic units execute instructions and periodically read and write data from and into memory. Fetching data from memory or writing data into memory can be slow operation, especially if data is not at the cache level. System operations often require accessing memory multiple times, which leads to creating several levels of memory hierarchy caches, where the closest level to the arithmetic and logic units has the shortest access time. Although cache memories reduce memory access time, there are still concerns that have to be addressed, such as cases where requested data is not loaded into cache memory, and thus such data has to be fetched from the lower levels within memory hierarchy, which is an expensive operation. In addition, in a multiprogramming environment a processor often changes the execution context which demands certain time to set up a new context [90].

In the dataflow paradigm, the source code is transformed into an execution graph where nodes in the graph are arithmetic or logic units [91]. Data streams are forwarded from memory into the dataflow engine, also referred to as DFE, where data is forwarded from one arithmetic or logic unit to another one. The dataflow engine contains a number of basic arithmetic and logic units, which are reconfigurable in an appropriate order depending on the execution graph. Each arithmetic unit can compute only one simple arithmetic operation that enables one to combine lots of units into one dataflow engine. During the execution process, an execution graph is configured before data arrives and then data actually flow through the pipelines of arithmetic and logic units all the way to the output.

In the control-flow paradigm, which is based on von–Neumann architecture, the compilation goes to the machine code level [89]. The control-flow paradigm can be referred to as computing in time because different operations are computed in different moments of time at the same functional units. In contrast, the dataflow paradigm can be referred to as computing in space because computations are placed dimensionally on a chip [89]. Speedwise, control-flow could compete with dataflow only if its implementation is based on special technologies like GaAs [92], and if a highly parallel computing paradigm is used [93]. Otherwise, dataflow is always powerwise superior.

4.4 Advantages of the dataflow paradigm

The dataflow paradigm demonstrates advantages in applications with massive workloads where power resources are limited, or in applications where power dissipation is just as important as execution speed. The dataflow paradigm can achieve significant energy efficient acceleration against the control-flow paradigm, but requires switching to a different architectural approach and a new programming model which is not trivial. In order to achieve acceleration, appropriate modifications in the source code need to be done and an execution graph has to be generated. In tensor operations, loops take most of the execution time; the goal is to migrate the execution of loops from the host machine to the accelerator, which implies migrating the execution of loops below the machine code level, from the software level to the hardware level.

In order to determine whether an algorithm is suitable for migrating to the dataflow paradigm, three main conditions have to be fulfilled. The first condition is related to data velocity and implies that an algorithm needs to perform complex computations with massive workloads in order to expect execution speed acceleration. The second condition is related to Amdahl's law. According to Amdahl's law, an algorithm can be divided into a serial parts and parallel parts. Parallel parts of an algorithm often consist of loops and in order to achieve acceleration, it is crucial that the parallel part takes 95% of the entire algorithm. Remarkable acceleration in terms of execution speed can be achieved only if serial parts continue to run on the control-flow machine and takes less than 5% of the application, and parallel parts migrate to the dataflow accelerator [19]. The third condition is related to the level of data reusability in the algorithm. Acceleration depends on the level of data reusability inside the migrated loops. Streaming data from the host machine to the dataflow accelerator is a relatively expensive operation, and it is not desirable to stream data if the level of reusability is not at a sufficient level. Only a high level of data reusability can outperform the negative influence of the slow interconnection bus between the host machine and the accelerator.

If the above-mentioned conditions are fulfilled, the level of acceleration that can be achieved depends on the optimization constructs used for an algorithm. The maximum performance can be achieved only if the following optimization constructs of the dataflow accelerator are exploited: (A) utilizing internal pipelines, (B) utilizing on-board (on-chip and off-chip) memories, and (C) reducing the precision of computations.

4.4.1 Utilizing internal pipelines

Pipelines in an execution graph represent connections between arithmetic and logic units. Each arithmetic or logic unit takes a certain amount of time to compute the result and in order to properly compute the result, input data has to arrive at the same time. In order to achieve the proper functioning of an execution graph, the compiler may insert data buffers on particular pipelines in order to meet deadlines for computations. If a pipeline contains a buffer, this means that it is not fully utilized and that there are moments when some units are waiting for input data in order to compute results. Utilizing internal pipelines is an optimization construct for smart utilization of idling situations by employing internal pipelines and avoiding data buffers. For example, when an execution graph is cyclic, the input of one pipeline is indirectly connected to the result of the same pipeline, so the execution graph needs to wait for the previous result in order to start computing the new result. There are several approaches in literature how to handle such cases by changing input data choreography [22].

4.4.2 On-chip and off-chip memories

The dataflow card has two types of memory on the board: on-chip memory, referred to as fast memory, and off-chip memory, referred to as large memory. Fast memory is located on the chip with terabytes/s of access bandwidth and has a small capacity of several megabytes. This type of memory is often used for storing small size data structures with a high level of reusability, such as meta results in loops. Unlike fast memory, large memory is off-chip but on-board memory with storage of several gigabytes and access bandwidth smaller than fast memory. Fast and large memories are placed near arithmetic and logic units so the data can be easily and efficiently retrieved from memory and saved in it [22]. The algorithms that exploit these memory advantages are able to achieve performance.

4.4.3 Low precision fixed-point computations

Dataflow FPGA-based accelerators enable custom number representation and low precision computations which have impact on overall performance. Switching from floating-point representation to fixed-point representation can improve performance of an algorithm by executing arithmetic and logic operations much faster. Each computational unit takes an amount of hardware resources on a chip. However, fixed-point representation allocates less hardware resources compared to floating-point representation. Also, if the numerical range for an algorithm can be determined, performance can be drastically improved by reducing number precision. For instance, if precision is reduced, the same results are obtained with acceptable tolerance of accuracy but several orders of magnitude faster. However, while reducing number precision and subsequently reducing allocated hardware resources, the performance can be drastically increased where precision is still acceptable [19].

5. Energy efficient implementations of tensor operations

This section proposes energy efficient implementations of 10 basic tensor operations commonly used in machine learning algorithms with dataflow accelerators. This section analyzes existing energy efficient implementations from open literature and then proposes dataflow-based implementation using dataflow accelerators. The goal of this section is to analyze which tensor operations are suitable for the dataflow paradigm under which conditions. All dataflow-based implementations exploit the advantages of the dataflow paradigm including arithmetic changes in operations, modifying input data choreography, utilizing internal pipelines, utilizing on-chip memory, and enabling low precision computations. Each implementation is compared against control-flow implementation through aspects such as execution speed, power dissipation, and complexity. Also, each implementation has several variants in order to explore the relation between optimization constructs and achieved performances.

As most machine learning problems use second-order tensors for computations, the following implementations are based on second-order tensors, with possibility to easily extend to higher order tensors for a specific use case.

This article doesn't evaluate performance on a specific dataset due to inability to find one dataset suitable for all operations. Also, the goal is to present general implementations which could be extended to higher order tensors without major changes. Tensor operations can be used on various levels of abstraction: data communications in noise [94], computational physics [95], or electronic business on the web [96]. If a control-flow supercomputer is to be tuned to tensor operations, its interconnection network can be optimized [97].

The following implementations utilize optimization constructs and it is analyzed how these optimizations influence execution time performance. The naive implementation, referred to as N in the evaluation figures, represents simple mapping of an algorithm or a part of the algorithm to the accelerator. Implementation with arithmetic changes, referred to as A in evaluation figures, represents the mapping of an algorithm or a part of algorithm with changes in basic arithmetic properties, such as association, distribution, and commutative property. In some cases, by changing and these basic arithmetic properties of an algorithm, execution time performance can be significantly improved, since some arithmetic operations are more suitable for dataflow accelerators than others. The implementation that involves internal pipelines and changes in input data choreography, also referred to as IP in evaluation figures, includes an optimization method where the host streams data in a specific order in which performance improvements may be achieved. Utilization of internal pipelines is often used when the execution graph contains a cyclic path. By utilizing the internal pipeline, latency is higher, but when the first result is computed, a result should be calculated in almost each tick. Implementations which utilize on-chip and off-chip memories, referred to as M in evaluation figures, reduce communication between the host program and the accelerator by storing data in on-board memories. These memories can improve execution speed performance due to efficient memory access bandwidth. Performance can be drastically improved due to lower allocation of resources by reducing number representation from the floating-point to the fixed-point number representation as well as precision. Such optimization construct is denoted as N in evaluation figures. Not all optimization constructs are suitable for all tensor operations. Different tensor operations may include different optimization constructs.

5.1 Tensor addition

Tensor addition represents a basic linear algebra operation which is used in almost any machine learning algorithm. Tensor addition is performed by adding the corresponding elements across axes. With tensors A and B, and their indices i and j, tensor addition can be formulated as shown in Eq. (1).

$$A_{ij} + B_{ij} = C_{ij} \tag{1}$$

Due to low implementational complexity, there are not many articles in open literature [81] that discuss energy efficient implementations of this operation. However, this implementation is important as a simple example of the dataflow paradigm, covering its basic concepts. In tensor addition, execution time is spent in loops and the entire algorithm can be migrated to the dataflow accelerator. In the naive implementation, the host program generates data and sends it in a rowwise order through interconnection to the dataflow card and waits for the result to be calculated. The dataflow compiler creates an execution graph by transforming loops into an execution graph. Each element of a tensor row has a separate pipeline that consists of only one arithmetic unit, which is the summation unit. By creating separate pipelines for elements in the execution graph, arithmetic units compute result elements simultaneously, which means that parts of a tensor row are computed separately and combined together before sending the computed row to the host program. If one iteration in the accelerator, referred to as a tick, takes one unit of time, it takes m units of time to compute the result of addition of two tensors with m rows. If input data choreography is changed, and instead of sending tensors in a rowwise order it sends entire tensors to the accelerator in one tick and it takes only one tick to compute the result. However, such an approach has limitations in hardware resources for high-order tensors and it depends on the hardware resources provided by the FPGA card. A part of the execution graph for tensor addition is shown in Fig. 1.

Sending data between the host program and the accelerator is an expensive operation and its cost can be minimized by utilizing dataflow memory instead. In cases where a set of the same data is repetitively used in the addition process, on-chip memory can be utilized for storing those data. By avoiding data movement between the host program and the accelerator and by storing data in the on-chip memory with data access bandwidth of TB/s, significant acceleration can be achieved. On-chip memory can store several MBs of data, which can be suitable for such simple operations. In that case, one tensor persists in on-chip memory, while the second tensor is streamed from the host program to the accelerator.

Furthermore, if the numerical range for the operation can be determined by switching from floating-point to fixed-point representation, the performance of the operation can be significantly increased. If precision goes down

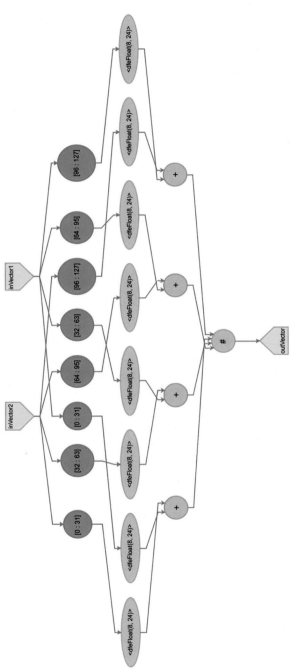

Fig. 1 Part of the execution graph that performs tensor addition. The execution graph contains four independent pipelines that perform addition of tensors. The last computational unit in the graph collects elements of a tensor and sends them back to the host program.

to the bit level, performance is enormously increased due to efficient bitwise execution of arithmetic and logic operations on FPGA cards. Fig. 2 shows how different optimization constructs affect performance in terms of execution time for tensor addition. If the host program sends entire tensors to the accelerator, computing the result is faster than iterating and computing the result in a rowwise order.

Each optimization construct method can be additionally improved by switching from floating-point to fixed-point number representation. Fig. 3 shows how reduced number precision affects execution speed. Beside the faster execution of an algorithm, when number precision is reduced, fewer hardware resources are allocated, and thus better parallelization can be done by creating more pipelines.

5.2 Tensor transpose

Tensor transpose is an arithmetic operation which changes the order of elements in a tensor. According to Eq. (2), symmetric elements over diagonals in a tensor exchange their positions.

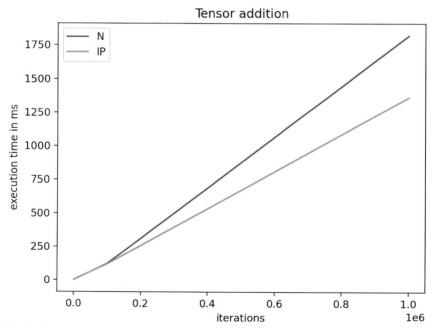

Fig. 2 Performance evaluation for tensor addition with optimization constructs.

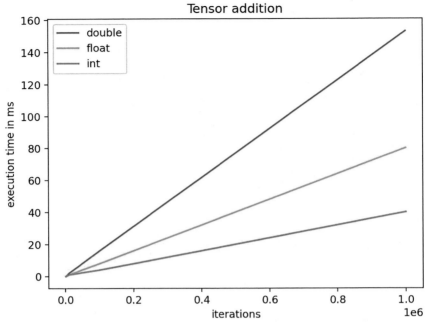

Fig. 3 Performance evaluation for tensor addition with number representation and precision.

$$\left[A^T\right]_{ij} = \left[\mathrm{A}\right]_{ji} \tag{2}$$

This operation can be efficiently used in machine learning for calculating the output of a layer in a neural network by transposing input data and multiplying them with corresponding weights. There are articles in open literature that introduce high-performance implementations of tensor transpose operations on powerful Intel and NVidia microprocessors [82]. However, there is no energy efficient implementation of this operation based on the dataflow architecture.

The nature of this iteration makes it suitable for implementation on the dataflow accelerator. In the naive implementation, the host program sends the entire tensor to the accelerator and waits to retrieve the result. This execution graph does not contain any arithmetic or logic units and it only describes the order of input and output pipelines. Such an approach is suitable for relatively small tensors that can be streamed to the accelerator in one tick. However, if a tensor cannot be streamed in one tick due to limited resources on the dataflow accelerator, the host program can send a tensor in a rowwise order and store the rows in on–chip memory. The memory based implementation therefore consists of two phases: (A) streaming a tensor to on–chip memory and (B) reading on–chip memory in the transposed order, as shown in Fig. 4. This implementation has a latency before the first result has been computed.

Another approach utilizes stream offsets where it is possible to fetch a neighbor in a stream which is suitable if there are not enough resources for storing tensor rows on a dataflow chip. Using such an approach, the host program sends elements of a tensor using a single stream. Using the stream offsets, the accelerator can dynamically calculate the position of the next element that should be streamed to the output and fetches it from the stream. Fig. 5 shows how different optimization constructs affect performance in

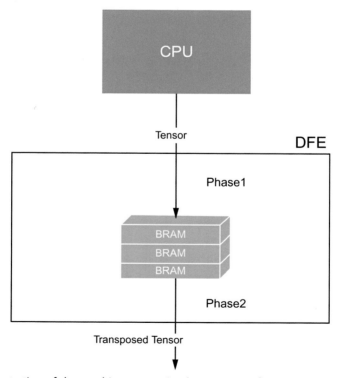

Fig. 4 Illustration of the on-chip memory implementation of tensor transpose, which consists of two phases. Phase 1 is streaming data to on-chip memory. Phase 2 is reading data from the memory in the transposed order.

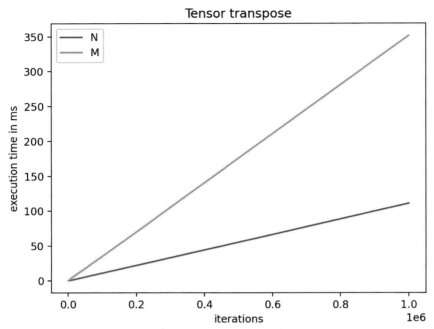

Fig. 5 Performance evaluation for tensor transpose with optimization constructs.

terms of execution time for tensor transpose operation. Although on-board memories reduce communication between the host program and the accelerator, it is necessary to hold the output until all parts of the tensor arrive to the accelerator.

5.3 Tensor product

A tensor product is a basic arithmetic operation in linear algebra, and as such has numerous applications in many areas. In machine learning, it is commonly used for calculating the output of a layer of a neural network. If A and B are tensors, their tensor product AB is a tensor calculated using Eq. (3).

$$t_{ij} = \sum_{k=1}^{m} a_{ik} b_{kj} \tag{3}$$

There are articles in open literature that introduce dataflow implementations of the tensor product [83,84]. Most of these articles present implementation details of the tensor product for a specific use case without discussing several optimization constructs and how they affect execution speed and power dissipation. In the proposed naive dataflow implementation, the dataflow accelerator receives two tensors and calculates a new tensor in parallel, where each element in a tensor has a separate pipeline. Pipelines from the first tensor are combined with pipelines from the second tensor, as shown in Fig. 6.

However, if a tensor cannot be streamed from the host program to the accelerator due to hardware limitations, the input data choreography can be modified in a way to tile input tensors into rows and columns. Using such an approach, the accelerator in the same tick receives a row of the first tensor and column of the second tensor and calculates the reduced sum.

The dataflow compiler contains several optimization constructs which are automatically applied to the execution graph before mapping onto the FPGA card [98]. The compiler optimizes the execution graph in order to produce the result in the minimal number of ticks. Tree reduction is an optimization construct that incurs a lower scheduling resource cost than a naive reduction and the operations are performed with lower latency. In naive implementations without compiler optimizations, the reduced sum is calculated sequentially. After applying the tree reduction, the dataflow accelerator computes an element of a reduced sum result using outputs from two neighbor pipelines from the previous step in the reduced sum.

Some use cases require the product between two tensors where one tensor is a constant or permanent for a number of iterations. On-board memory-based implementation is suitable for such use cases where one tensor is stored in on-chip memory, while a set of other tensors are streamed from the host program to the accelerator. When a tensor arrives to the accelerator, one tensor is fetched from on-chip memory while another one is already at the input of arithmetic units. Such an approach is also efficient when the algorithm works with a small set of tensors that can be stored in on-chip memory. Otherwise, if a set of tensors cannot be stored in the memory, the context of on-chip memory has to be changed periodically due to capacity limits.

The tensor product can be compute-intensive operation for a massive data workloads. An example would be training a neural network with a number of layers and neurons on those layers. Fig. 7 shows execution speeds of naive tensor product implementation on the dataflow and control-flow paradigms. Naive dataflow implementation achieves better performance that the control-flow implementation with significantly lower power dissipation.

Figs. 8–10 show how different number precisions influence the allocation of hardware resources on the FPGA card. From switching from floating-point number representation to reduced fixed-point number representation, the execution graph for the tensor product can perform better parallelization by allocating more hardware resources.

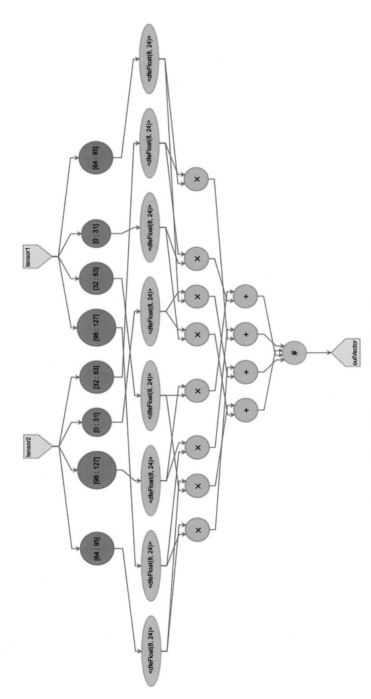

Fig. 6 Execution graph for tensor product with four different pipelines for each tensor.

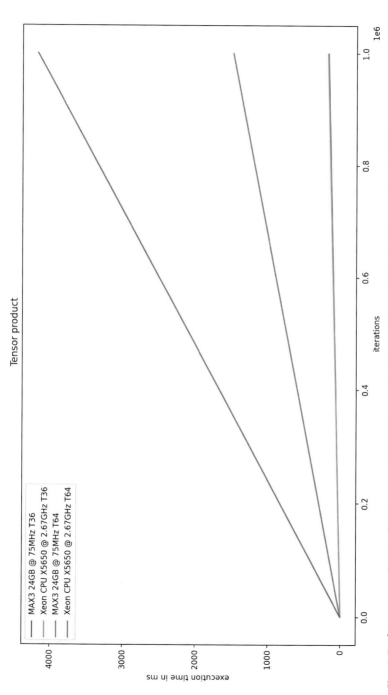

Fig. 7 Performance evaluation for tensor product between dataflow and control-flow implementations for different tensor sizes. The dataflow accelerator MAX3 24GB @ 75 MHz is compared with Intel Xeon CPU X5650 @ 2.67 GHz.

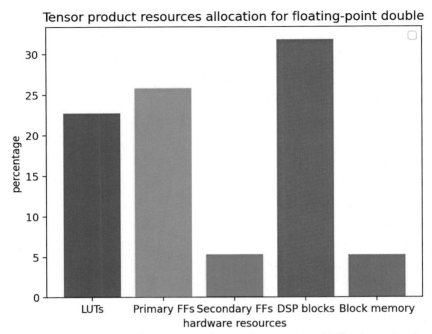

Fig. 8 Hardware allocation for tensor product implementation with floating-point double precision.

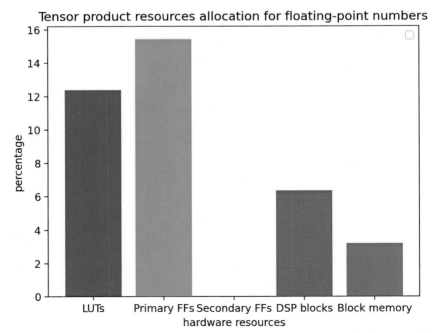

Fig. 9 Hardware allocation for tensor product implementation with floating-point precision.

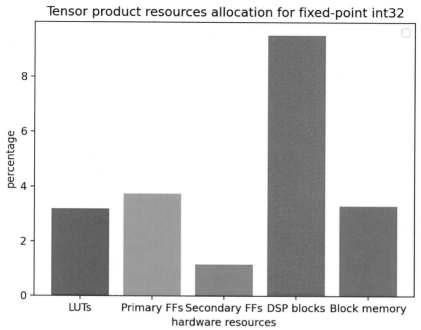

Fig. 10 Hardware allocation for tensor product implementation with fixed-point int32 precision.

5.4 Divergence, curl, and gradient

From a computational perspective, divergence, curl, gradient, and gradient descent methods can be seen as a tensor product, which has already been discussed in the previous section. In some cases, these could be represented as special cases of the tensor product, while calculating the divergence can be considered as the product of two tensors with different orders, where for example, one tensor can be a first-order tensor while another one can be a second-order tensor. In the naive dataflow implementation, the host program sends tensors to the accelerator, which has already been discussed in previous sections. Such an approach is suitable for tensors that cannot be stored in on–chip memory due its capacity.

However, if a first-order tensor can be stored in on–chip memory with high access bandwidth, remarkable performance can be achieved by streaming only 1 s-order tensor to the accelerator. A new tensor is computed in each tick of the accelerator by utilizing parallel internal pipelines, as shown in Fig. 11.

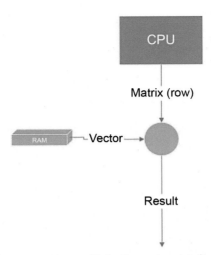

Fig. 11 Illustration of tensor-vector multiplication using dataflow accelerator.

This subsection presents the last one of the basic arithmetic operations and their implementations on dataflow accelerators. It is shown that even for such simple operations acceleration is achieved using the dataflow paradigm. The overall performance evaluation is discussed in the following section, while the following subsections present the implementation details of complex tensor operations and their optimization constructs.

5.5 Tensor inverse

The importance of tensor inverse for simultaneously solving linear equations is crucial in statistics and machine learning. In linear algebra, tensor A is invertible if there is another tensor A^{-1} that satisfies Eq. (4). A square tensor that is not invertible is called a singular tensor where its determinant is equal to 0. Tensor inverse in an operation that requires a complex computation model, and could be efficiently computed using one of the factorization methods.

$$AA^{-1} = A^{-1}A = I \qquad (4)$$

There are articles in open literature that introduce dataflow implementations of the tensor product [85,86]. Most of these articles present implementation details of tensor inverse for a specific use case without discussing certain optimization constructs and how those affect execution speed and power dissipation. This subsection sheds light on the following factorization methods that can be used for calculating the inversion of a tensor: (A) Doolittle algorithm, (B) Crout decomposition, and (C) Cholesky decomposition. The first two methods are similar to LU decomposition with some arithmetic changes, and this subsection proposes the implementation details of LU and Cholesky decompositions.

LU decomposition refers to the factorization of a tensor with proper row and column permutations into two factors: lower triangular tensor L and an upper triangular tensor U. Cholesky decomposition transforms a tensor into the product of a lower triangular tensor and its conjugate transpose. Cholesky decomposition can be computed only for positive-definite tensors. The naive dataflow implementations of the above-mentioned decompositions are based on data streaming where in each iteration the host program sends the entire tensor to the dataflow accelerator. Dataflow implementations can be divided into two phases: (A) LU/Cholesky decomposition and (B) computing the inverse of a tensor using the result from the previous phase.

In the first phase, each element in a tensor has its own separate pipeline. These pipelines are created by unrolling the control-flow loops during compilation time. One pipeline is connected with other pipelines in order to use previously computed elements of a new decomposed tensor. Before the first result is computed, there is a latency resulting from the depth of the execution graph. At the final stage of the execution graph, all elements are collected in result tensors.

In the proposed solution, splitting factorization methods into two phases allows for full internal pipeline utilization. Once the first phase is finished, data are forwarded to the second phase for computing the inverse of a tensor, while at the same time the first phase can be performed for a new tensor. Furthermore, the performance of the operation can be additionally improved by switching to low-precision number representation.

5.6 Primary invariants

The invariants of a tensor are the coefficients of the characteristic polynomial of tensor A that satisfies Eq. (5), where E is the identity tensor and λ is the polynomial's indeterminate.

$$p(\lambda) = det(A - \lambda I) \tag{5}$$

The first invariant, known as trace, is the sum of diagonal components, as shown in Eq. (6). The n-th invariant is the determinant of the tensor (det A).

$$I_A = A_{11} + A_{22} + \ldots + A_{nn} = trace(A) \tag{6}$$

In the existing open literature there are no dataflow implementations of primary and principal invariants which have direct impact in machine learning algorithms, but since they are an important linear algebra concept, this section presents their dataflow implementation. The best approach for finding the invariants and determinant of a tensor is using the dataflow paradigm is one of the factorization methods which have already been discussed in the previous subsection. The proposed dataflow implementation computes the invariants and determinant of a tensor as a scaling factor of the transformation described by the tensor. In the proposed implementation, only the second phase differs from the previous subsection.

5.7 Principal invariants

As indicated in the Cayley-Hamilton theorem and presented in Eq. (7), an invertible tensor with non-zero determinant can thus be written as the n-th order polynomial expression. The proposed dataflow implementation computes characteristic polynomial coefficients using the factorization method called power iterations.

$$p(A) = A^n + c_{n-1}A^{n-1} + \ldots c_1 A + (-1)^n det(A)I = 0 \tag{7}$$

The power iteration algorithm is an eigenvalue algorithm that computes the greatest eigenvalue and corresponding eigenvector, which is described in Eq. (8). The main disadvantage of such an approach is that it slowly converges to the result. This algorithm starts with vector $b0$, which may be an approximation to the eigenvector, where in every iteration vector b_k is multiplied with tensor A and then normalized.

$$b_{k+1} = \frac{Ab_k}{\|Ab_k\|} \tag{8}$$

The proposed dataflow implementation of the introduced algorithm exploits the advantage of off-chip memory. The tensor is streamed to off-chip memory, which has a higher bandwidth than the link between the host program and the dataflow accelerator. In each iteration, the dataflow card retrieves data from on-board memory and computes a new approximation of the eigenvalue and eigenvector, after which it transfers the result back to the memory. At the end when the approximation is computed, the dataflow card sends the result to the host machine.

5.8 Eigenvalues and eigenvectors

The concept of eigenvalues and eigenvectors naturally extends to arbitrary linear transformations on arbitrary vector spaces. In Eq. (9), T is a linear transformation from vector space V and v is an eigenvector of T, if $T(v)$ is a scalar multiple of v.

$$T(v) = \lambda v \tag{9}$$

There are a number of articles in open literature which propose dataflow implementations of eigenvalues and eigenvectors used for dimensionality reduction in machine learning [87,88]. Calculating eigenvalues and eigenvectors for a tensor is not a trivial problem, especially for high-order tensors. The power iteration algorithm from the previous subsection can be utilized for calculating eigenvalues and eigenvectors of a tensor. In this subsection, the proposed implementation of eigenvalues and eigenvectors is based on QR decomposition.

Any real square matrix may be decomposed as QR, where Q is an orthogonal matrix and R is an upper triangular matrix. The columns in the Q matrix represent eigenvectors, while the diagonal elements in the R matrix present eigenvalues. The proposed dataflow solution implements the QR decomposition using two different algorithms: (A) Gram-Schmidt algorithm and (B) Householder algorithm.

The Householder algorithm can be expressed as a transformation that takes a tensor and reflects it about some plane or hyperplane. This algorithm performs in-place computations, which allocates on-chip memory for storing results between iterations. The Gram-Schmidt algorithm takes a finite, linearly independent set and generates an orthogonal set that spans the same k-dimensional subspace. Due to rounding errors, this process does not always produce an orthogonal matrix. This is an iterative process suitable for the dataflow accelerator and on-board memory.

In dataflow implementations, the accelerator receives the entire tensor where elements have their own separate pipelines. The data dependencies of execution graphs are acyclic, which means that internal pipelines are fully utilized without data buffering between arithmetic or logic units.

5.9 Spectral decomposition

In linear algebra, spectral decomposition, or sometimes eigendecomposition is the factorization of a tensor into a canonical form where a tensor is represented in terms of its eigenvalues and eigenvectors. Using the eigenvalues and eigenvectors of a tensor makes it easy to calculate other tensor properties, such as singular values, rank, pseudo-inverse, and linear differential equations. The previous subsection presents the proposed dataflow implementations of the Gram-Schmidt and Householder algorithms for computing eigenvalues and eigenvectors. This subsection introduces the dataflow implementation of the Jacobi eigen algorithm which is also used for computing eigenvalues and eigenvectors. The Jacobi eigen algorithm is an iterative method based on rotations and can be applied only on real symmetric tensors.

Each Jacobi rotation can be done in n steps when the pivot element p is known. However, the search for p requires an inspection of all off-diagonal elements. In the proposed dataflow implementation, the host program sends data to off-chip memory. In each iteration, data are retrieved from off-chip memory, rotations are computed, and data are transferred back to memory. The same implementation pattern is also used in previous tensor operations. Once when an eigenvector is computed, the accelerator fetches the result from on-board memory and transfers it to the host machine.

5.10 Tensor rank

Tensor rank is used in machine learning for finding a tensor of a certain rank which approximates a given tensor. Gaussian elimination is an algorithm used for solving systems of linear equations and can be used for calculating the rank of a tensor using an echelon form. The process of row reduction makes use of elementary row operations, and can be divided into two parts. The first part is forward elimination which reduces a given tensor to a row echelon form, while the second part is the back substitution which continues to use row operations over the reduced form until the solution is found. In other words, the algorithm puts the tensor into a reduced row echelon form, which is important for calculating the rank of a tensor. In the proposed dataflow implementation, row swapping in memory is simulated with reordering internal pipelines. The proposed implementation means computing the rank of a tensor simultaneously in a rowwise order. Execution speed depends on the size of a tensor where high parallelization speedup can be achieved.

6. Performance evaluation

This section discusses the performance evaluation of 10 basic tensor operations on dataflow accelerators and compares them against control-flow implementations across different aspects such as execution speed, power dissipation, complexity, and meantime between failure [99]. These 10 tensor operations are chosen in a way to cover all important algorithms used in machine learning, where some of them are complex operations with massive workloads.

6.1 Execution speed

Fig. 12 shows speedup which is achieved for 10 tensor operations using dataflow accelerators. Complex tensor operations such as tensor decompositions are able to exploit optimization constructs of dataflow accelerators

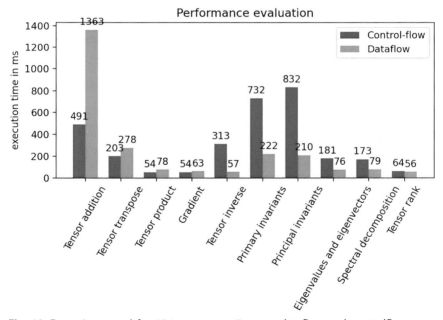

Fig. 12 Execution speed for 10 tensor operations on dataflow and controlflow paradigms in ms. Results are obtained for different data sets, depending on available resources on the dataflow accelerator. Dataflow accelerator MAX3 24GB @ 75 MHz is compared with Intel Xeon CPU X5650 @ 2.67 GHz.

and achieve significant speedup compared against conventional control-flow implementations. However, simple tensor operations like addition and transpose, which do not have complex implementations, do not achieve speedup so the controlflow paradigm is therefore better. A direct comparison of different architectural approaches such as dataflow and control-flow is not possible. In order to compare speedup per watt and conclude which tensor operations are energy efficient, this comparison takes the number of transistors in direct relation to power dissipation as a relevant scaling factor between the two paradigms.

6.2 Power dissipation

Power dissipation depends on the clock frequency and number of transistors. Control-flow microprocessors work on GHzs, while dataflow accelerators work on MHzs. Using the dataflow paradigm, power dissipation is more than $20 \times$ lower compared to the conventional control-flow paradigm. Lower dissipation of the dataflow paradigm also stems from the fact that transistors dissipate during a shorter period of time since applications get executed faster. Fig. 13 shows speedup per W and energy efficiency for 10 tensor operations on dataflow accelerators against control-flow implementations.

6.3 Complexity

The complexity of two paradigms is measured with the number of transistors. For example, control-flow microprocessors such as GT200 Tesla or Intel Xeon have about 2,000,000,000 transistors, while dataflow accelerators such as Xilinx or Altera have roughly about 200,000,000 transistors. In other words, the dataflow paradigm can perform faster computation with 10 times fewer transistors and a smaller clock frequency.

6.4 Mean time between failures

The mean time between failure domains depends a lot on the transistor count, power dissipation, and presence of components prone to failure. If a microprocessor works at higher clock frequency, the probability of failure is also higher. One of the greatest challenges in high-performance computing is avoiding failures and recovering from them. Clusters of accelerators require a simultaneous use of hundreds of thousands of processing, storage, and networking elements. With a large number of elements involved, element failure will be frequent. The extensive analysis of failures in petascale computers is presented in article [100].

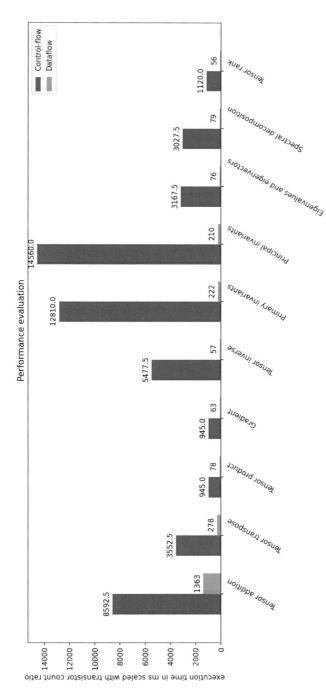

Fig. 13 Execution speed per W for 10 tensor operations on dataflow and control-flow paradigms per W and transistor count in ms. Results shown in the table are obtained for different data sets, depending on available resources on the dataflow accelerator. The dataflow accelerator MAX3 24GB @ 75 MHz is compared with Intel Xeon CPU X5650 @ 2.67 GHz. The ratio between control-flow and dataflow transistor count is 17.5.

7. Conclusion

Since this article analyzed implementations in two different domains, four sets of conclusions can be derived. They are related to performance, power, complexity, and mean time between failures.

In the performance domain, the main conclusion is that the relative ratio of the four compared architectures depends a lot on the structure of the loops and data involved. The loop characteristics that favorize dataflow are: an extremely high contribution of loops to overall execution time, an extremely high level of data reusability within loops. The data characteristics that favorize dataflow are: an extremely high data volume and the streaming nature of data input. In the power domain, the main conclusion is that the relative ratio of the four compared architectures depends a lot on the clock frequency and the amount of resources that consume power. The dataflow paradigm is characterized by: a slower clock, and no von-Neumann resources known for relatively high power dissipation due to a relatively high number of transistors needed for implementation, such as instruction predictors, data predictors, caches, and memory management. Finally, the lower dissipation of the dataflow paradigm also comes from the fact that transistors dissipate during a shorter period of time since applications get executed faster. Also, architectures that are more uniform in their structure may be implementable with a lower voltage, which means less dissipation.

In the complexity domain, the main conclusion is that the relative ratio of the four compared architectures depends a lot on: the transistor count needed for the implementation of the paradigm. Again, paradigms needing fewer transistors for implementation are characterized by lower complexity. Other factors, not included into this study, which affect the complexity of implementation are: uniformity of structure, implement ability with simple structure elements, and lower cooling complexity due to lower power dissipation. This study relies on the transistor count data given by the manufacturers.

In the mean time between failure domains, the main conclusion is that the relative ratio of the four compared architectures depends a lot on: the transistor count, power dissipation, presence of components prone to failure, and implementation technology issues. This study relies on the mean time between failure data provided by manufacturers.

All the above-mentioned issues interact synergistically. If an approach is based on two or more paradigms working in symbiosis, the relative contributions coming from each involved paradigm have to be superimposed in a selective manner (differently for different applications, in accordance with how much an application spends time in a each and every paradigm involved). These aspects (synergy, symbiosis, superimposing, and selectivity) will be the subjects of a follow-up research.

Obtained results show that the dataflow implementations of tensor operations can achieve remarkable performance in terms of speed per watt against control-flow implementations and thus could be used in embedded systems where power resources are limited. This article is important for further research in energy efficient implementations of machine learning algorithms which is extremely important with proliferation of data and devices.

References

[1] A. Oussous, F.-Z. Benjelloun, A.A. Lahcen, S. Belfkih, Big data technologies: a survey, J. King Saud Univ. Comp. Info. Sci. 30 (4) (2018) 431–448.

[2] W.A. Günther, M.H.R. Mehrizi, M. Huysman, F. Feldberg, Debating big data: a literature review on realizing value from big data, J. Strateg. Inf. Syst. 26 (3) (2017) 191–209.

[3] S. Wolfert, L. Ge, C. Verdouw, M.-J. Bogaardt, Big data in smart farming: a review, Agr. Syst. 153 (2017) 69–80.

[4] Y. Wang, L. Kung, T.A. Byrd, Big data analytics: understanding its capabilities and potential benefits for healthcare organizations, Technol. Forecasting Social Change 126 (2018) 3–13.

[5] I. Yaqoob, I.A.T. Hashem, A. Gani, S. Mokhtar, E. Ahmed, N.B. Anuar, A.V. Vasilakos, Big data: from beginning to future, Int. J. Inf. Manag. 36 (6) (2016) 1231–1247.

[6] L. Jollans, R. Boyle, E. Artiges, T. Banaschewski, S. Desrivières, A. Grigis, J.-L. Martinot, T. Paus, M.N. Smolka, H. Walter, et al., Quantifying performance of machine learning methods for neuroimaging data, Neuroimage 199 (2019) 351–365.

[7] M. Kotlar, D. Bojić, M. Punt, V. Milutinović, Survey of deployment locations and underlying hardware architectures for contemporary deep neural networks, Int. J, Distrib. Sens. Netw. 15 (8) (2019).

[8] A. Mayr, D. Kißkalt, M. Meiners, B. Lutz, F. Schäfer, R. Seidel, A. Selmaier, J. Fuchs, M. Metzner, A. Blank, et al., Machine learning in production: potentials, challenges and exemplary applications, Proc. CIRP 86 (2019) 49–54.

[9] R. Vinayakumar, K. Soman, P. Poornachandran, V.K. Menon, A deep-dive on machine learning for cyber security use cases, in: Machine Learning for Computer and Cyber Security, CRC Press, 2019, pp. 122–158.

[10] S. Emerson, R. Kennedy, L. O'Shea, J. O'Brien, Trends and applications of machine learning in quantitative finance, in: 8th International Conference on Economics and Finance Research (ICEFR 2019), 2019.

[11] G. Guan, W. Dong, Y. Gao, K. Fu, Z. Cheng, Tinylink: a holistic system for rapid development of IoT applications, in: Proceedings of the 23rd Annual International Conference on Mobile Computing and Networking, 2017, pp. 383–395.

[12] A. Einstein, Relativity: The Special and the General Theory, Princeton University Press, 2015.

[13] L. Hardesty, Faster Big-Data Analysis, MIT News, 2017.

[14] Q. Huang, P. Smolensky, X. He, L. Deng, D. Wu, Tensor product generation networks for deep NLP modeling, arXiv (2017). preprint arXiv:1709.09118.

[15] A. Tjandra, S. Sakti, S. Nakamura, Compressing recurrent neural network with tensor train, in: 2017 International Joint Conference on Neural Networks (IJCNN), IEEE, 2017, pp. 4451–4458.

[16] M. Abadi, P. Barham, J. Chen, Z. Chen, A. Davis, J. Dean, M. Devin, S. Ghemawat, G. Irving, M. Isard, et al., Tensorflow: A system for large-scale machine learning, in: 12th {USENIX} symposium on operating systems design and implementation ({OSDI} 16), 2016, pp. 265–283.

[17] D. Tuffley, Google's Release of Tensorflow Could Be a Gamechanger in the Future of AI, The conversation, 2015.

[18] S. Cass, Taking AI to the edge: Google's TPU now comes in a maker-friendly package, IEEE Spectr. 56 (5) (2019) 16–17.

[19] M.J. Flynn, O. Mencer, V. Milutinovic, G. Rakocevic, P. Stenstrom, R. Trobec, M. Valero, Moving from petaflops to petadata, Commun. ACM 56 (5) (2013) 39–42.

[20] D. Milutinovic, V. Milutinovic, B. Soucek, The Honeycomb Architecture, Proceedings of the IEEE, 1987.

[21] V. Milutinovic, M. Kotlar, M. Stojanovic, I. Dundic, N. Trifunovic, Z. Babovic, DataFlow Supercomputing Essentials, Springer, 2017.

[22] N. Trifunovic, V. Milutinovic, J. Salom, A. Kos, Paradigm shift in big data super-computing: dataflow vs. controlflow, J. Big Data 2 (1) (2015) 4.

[23] J.-S. Chou, T.-K. Nguyen, Forward forecast of stock price using sliding-window metaheuristic-optimized machine-learning regression, IEEE Trans. Ind. Inf. 14 (7) (2018) 3132–3142.

[24] A. Rajkomar, J. Dean, I. Kohane, Machine learning in medicine, N. Engl. J. Med. 380 (14) (2019) 1347–1358.

[25] U. Fiore, A. De Santis, F. Perla, P. Zanetti, F. Palmieri, Using generative adversarial networks for improving classification effectiveness in credit card fraud detection, Inform. Sci. 479 (2019) 448–455.

[26] M. Hasan, M.M. Islam, M.I.I. Zarif, M. Hashem, Attack and anomaly detection in IoT sensors in IoT sites using machine learning approaches, Internet Things 7 (2019).

[27] J. Chaki, N. Dey, Pattern analysis of genetics and genomics: a survey of the state-of-art, Multimed. Tools Appl. 79 (2019) 1–32.

[28] R. Cuocolo, M.B. Cipullo, A. Stanzione, L. Ugga, V. Romeo, L. Radice, A. Brunetti, M. Imbriaco, Machine learning applications in prostate cancer magnetic resonance imaging, Eur. Radiol. Exp. 3 (1) (2019) 1–8.

[29] B. Sekeroglu, K. Dimililer, K. Tuncal, Student performance prediction and classifica-tion using machine learning algorithms, in: Proceedings of the 2019 8th International Conference on Educational and Information Technology, 2019, pp. 7–11.

[30] C.L. Stergiou, A.P. Plageras, K.E. Psannis, B.B. Gupta, Secure machine learning sce-nario from big data in cloud computing via internet of things network, in: Handbook of Computer Networks and Cyber Security, Springer, 2020, pp. 525–554.

[31] X. Xu, Q. Liu, Y. Luo, K. Peng, X. Zhang, S. Meng, L. Qi, A computation offloading method over big data for IoT-enabled cloud-edge computing, Futur. Gener. Comput. Syst. 95 (2019) 522–533.

[32] R.S. Williams, What's next?[the end of moore's law], Comput. Sci. Eng. 19 (2) (2017) 7–13.

[33] M.M. Waldrop, More than moore, Nature 530 (7589) (2016) 144–148.

[34] T. Dobravec, P. Bulić, Comparing CPU and GPU implementations of a simple matrix multiplication algorithm, Int. J. Comput. Electr. Eng. 9 (2) (2017) 430–438.

[35] A. Reuther, P. Michaleas, M. Jones, V. Gadepally, S. Samsi, J. Kepner, Survey and benchmarking of machine learning accelerators, arXiv (2019). preprint arXiv:1908.11348.

[36] F. Al-Turjman, I. Baali, Machine learning for wearable IoT-based applications: a survey, Trans. Emerg. Telecommun. Technol. (2019) e3635.

[37] E. Nurvitadhi, G. Venkatesh, J. Sim, D. Marr, R. Huang, J. Ong Gee Hock, Y.T. Liew, K. Srivatsan, D. Moss, S. Subhaschandra, et al., Can FPGAs beat GPUs in accelerating next-generation deep neural networks? in: *Proceedings of the 2017 ACM/SIGDA International Symposium on Field-Programmable Gate Arrays*, ACM, 2017, pp. 5–14.

[38] C. Baskin, N. Liss, A. Mendelson, E. Zheltonozhskii, Streaming architecture for large-scale quantized neural networks on an FPGA-based dataflow platform, arXiv (2017). *preprint arXiv:1708.00052*.

[39] N. Voss, M. Bacis, O. Mencer, G. Gaydadjiev, W. Luk, Convolutional Neural Networks on Dataflow Engines, in: *2017 IEEE International Conference on Computer Design (ICCD)*, IEEE, 2017, pp. 435–438.

[40] M.B. Taylor, L. Vega, M. Khazraee, I. Magaki, S. Davidson, D. Richmond, ASIC clouds: specializing the datacenter for planet-scale applications, Commun. ACM 63 (7) (2020) 103–109.

[41] Y. Zhou, H. Ren, Y. Zhang, B. Keller, B. Khailany, Z. Zhang, Primal: power inference using machine learning, in: *Proceedings of the 56th Annual Design Automation Conference 2019*, 2019, pp. 1–6.

[42] K.-C. Wu, Y.-W. Tsai, Structured ASIC, evolution or revolution? in: *Proceedings of the 2004 International Symposium on Physical Design*, ACM, 2004, pp. 103–106.

[43] A.K. Sen, P.K. Tiwari, Security issues and solutions in cloud computing, IOSR J. Comput. Eng. 19 (2) (2017) 67–72.

[44] D. Danielson, Vectors and Tensors in Engineering and Physics, CRC Press, 2018.

[45] G. Strang, Linear Algebra and Learning From Data, Wellesley-Cambridge Press, 2019.

[46] L. Liţă, E. Pelican, A low-rank tensor-based algorithm for face recognition, App. Math. Model. 39 (3–4) (2015) 1266–1274.

[47] N.D. Sidiropoulos, L. De Lathauwer, X. Fu, K. Huang, E.E. Papalexakis, C. Faloutsos, Tensor decomposition for signal processing and machine learning, IEEE Trans. Signal Process. 65 (13) (2017) 3551–3582.

[48] S. Rabanser, O. Shchur, S. Günnemann, Introduction to tensor decompositions and their applications in machine learning, arXiv (2017). *preprint arXiv:1711.10781*.

[49] A. Jaffe, R. Weiss, B. Nadler, S. Carmi, Y. Kluger, Learning binary latent variable models: a tensor eigenpair approach, in: *International Conference on Machine Learning*, 2018, pp. 2196–2205.

[50] J. Virta, B. Li, K. Nordhausen, H. Oja, Independent component analysis for tensor-valued data, J. Multivar. Anal. 162 (2017) 172–192.

[51] C.J. Hillar, L.-H. Lim, Most tensor problems are NP-hard, J. ACM 60 (6) (2013) 45.

[52] J. Li, J. Sun, R. Vuduc, HICOO: hierarchical storage of sparse tensors, in: SC18: International Conference for High Performance Computing, Networking, Storage and Analysis, IEEE, 2018, pp. 238–252.

[53] K. Hegde, H. Asghari-Moghaddam, M. Pellauer, N. Crago, A. Jaleel, E. Solomonik, J. Emer, C.W. Fletcher, Extensor: An accelerator for sparse tensor algebra, in: *Proceedings of the 52nd Annual IEEE/ACM International Symposium on Microarchitecture*, 2019, pp. 319–333.

[54] H. Wang, W. Liu, K. Hou, W.-c. Feng, Parallel transposition of sparse data structures, in: *Proceedings of the 2016 International Conference on Supercomputing*, 2016, pp. 1–13.

[55] D.A. Matthews, High-performance tensor contraction without transposition, SIAM J. Sci. Comput. 40 (1) (2018) C1–C24.

[56] I. Goodfellow, Y. Bengio, A. Courville, Deep feedforward networks, Deep Learn. (2016) 168–227.

[57] K. Wan, H. Sun, M. Ji, D. Tuninetti, G. Caire, Cache-aided matrix multiplication retrieval, arXiv (2020). preprint arXiv:2007.00856.

[58] M. Cenk, M.A. Hasan, On the arithmetic complexity of strassen-like matrix multiplications, J. Symb. Comput. 80 (2017) 484–501.

[59] J. Huang, T.M. Smith, G.M. Henry, R.A. van de Geijn, Strassen's algorithm reloaded, in: *SC'16: Proceedings of the International Conference for High Performance Computing, Networking, Storage and Analysis*, IEEE, 2016, pp. 690–701.

[60] F.L. Gall, F. Urrutia, Improved rectangular matrix multiplication using powers of the coppersmith-winograd tensor, in: *Proceedings of the Twenty-Ninth Annual ACM-SIAM Symposium on Discrete Algorithms*, SIAM, 2018, pp. 1029–1046.

[61] M. Al-Mouhamed, A. Fatayer, N. Mohammad, et al., Optimizing the matrix multiplication using strassen and winograd algorithms with limited recursions on many-core, Int. J. Parallel Program. 44 (4) (2016) 801–830.

[62] M. Muller, S. Guha, E.P. Baumer, D. Mimno, N.S. Shami, Machine learning and grounded theory method: convergence, divergence, and combination, in: *Proceedings of the 19th International Conference on Supporting Group Work*, 2016, pp. 3–8.

[63] X. Lian, W. Zhang, C. Zhang, J. Liu, Asynchronous decentralized parallel stochastic gradient descent, in: *International Conference on Machine Learning*, 2018, pp. 3043–3052.

[64] V. Citro, P. Luchini, F. Giannetti, F. Auteri, Efficient stabilization and acceleration of numerical simulation of fluid flows by residual recombination, J. Comput. Phys. 344 (2017) 234–246.

[65] J. Hu, S. Gan, High order method for black–scholes PDE, Comput. Math. Appl. 75 (7) (2018) 2259–2270.

[66] G. Raskutti, M.W. Mahoney, A statistical perspective on randomized sketching for ordinary least-squares, J. Mach. Learn. Res. 17 (1) (2016) 7508–7538.

[67] T.A. Smaglichenko, A.V. Smaglichenko, W.R. Jacoby, M.K. Sayankina, Cluster algorithm integrated with modification of Gaussian elimination to solve a system of linear equations, in: *Artificial Intelligence and Evolutionary Computations in Engineering Systems*, Springer, 2020, pp. 583–591.

[68] G. Shabat, Y. Shmueli, Y. Aizenbud, A. Averbuch, Randomized LU decomposition, Appl. Comput. Harmon. Anal. 44 (2) (2018) 246–272.

[69] R.A. Tapia, J.E. Dennis Jr., J.P. Schaüfermeyer, Inverse, shifted inverse, and Rayleigh quotient iteration as newton's method, Siam Rev. 60 (1) (2018) 3–55.

[70] T. Kaczorek, Cayley-hamilton theorem for drazin inverse matrix and standard inverse matrices, Bull. Pol. Acad. Sci. Tech. Sci. 64 (2016).

[71] B. Jiang, J. Zhang, Least-squares migration with a blockwise hessian matrix: a prestack time-migration approach, Geophysics 84 (4) (2019) R625–R640.

[72] M. Drass, J. Schneider, S. Kolling, Novel volumetric helmholtz free energy function accounting for isotropic cavitation at finite strains, Mater. Des. 138 (2018) 71–89.

[73] Q. Liu, F. Davoine, J. Yang, Y. Cui, Z. Jin, F. Han, A fast and accurate matrix completion method based on QR decomposition and $L_{2,1}$-norm minimization, IEEE Trans. Neural Netw. Learn. Syst. 30 (3) (2018) 803–817.

[74] E.D. Nino-Ruiz, A. Sandu, X. Deng, A parallel implementation of the ensemble kalman filter based on modified cholesky decomposition, J. Comput. Sci. 36 (2019) 100654.

[75] J. Lever, M. Krzywinski, N. Altman, Points of Significance: Principal Component Analysis, Nature Publishing Group, 2017.

[76] M.U. Torun, O. Yilmaz, A.N. Akansu, FPGA, GPU, and CPU implementations of jacobi algorithm for eigenanalysis, J. Parallel Distrib. Comput. 96 (2016) 172–180.

[77] J. Vogel, J. Xia, S. Cauley, V. Balakrishnan, Superfast divide-and-conquer method and perturbation analysis for structured eigenvalue solutions, SIAM J. Sci. Comput. 38 (3) (2016) A1358–A1382.

[78] S. Padhy, L. Sharma, S. Dandapat, Multilead ECG data compression using SVD in multiresolution domain, Biomed. Signal Process. Control 23 (2016) 10–18.

[79] R. Zhang, A. Shapiro, Y. Xie, in: Statistical Rank Selection for Incomplete Low-Rank Matrices, ICASSP 2019–2019 IEEE International Conference on Acoustics, Speech and Signal Processing (ICASSP), IEEE, 2019, pp. 2912–2916.

[80] Y. Nakatsukasa, T. Soma, A. Uschmajew, Finding a low-rank basis in a matrix subspace, Math. Program. 162 (1–2) (2017) 325–361.

[81] T.B. Preußer, Generic and universal parallel matrix summation with a flexible compression goal for Xilinx FPGAs, in: 2017 27th International Conference on Field Programmable Logic and Applications (FPL), IEEE, 2017, pp. 1–7.

[82] D.I. Lyakh, An efficient tensor transpose algorithm for multicore CPU, Intel Xeon Phi, and NVidia Tesla GPU, Comput. Phys. Commun. 189 (2015) 84–91.

[83] L. Zhuo, V.K. Prasanna, High-performance and parameterized matrix factorization on FPGAs, in: 2006 International Conference on Field Programmable Logic and Applications, IEEE, 2006, pp. 1–6.

[84] S. Belkacemi, K. Benkrid, D. Crookes, A. Benkrid, Design and implementation of a high performance matrix multiplier core for Xilinx Virtex FPGAs, in: *2003 IEEE International Workshop on Computer Architectures for Machine Perception*, IEEE, 2003, p. 4.

[85] A. Hadizadeh, M. Hashemi, M. Labbaf, M. Parniani, A matrix-inversion technique for FPGA-based real-time EMT simulation of power converters, IEEE Trans. Ind. Electron. 66 (2) (2018) 1224–1234.

[86] Y. Xu, D. Li, Y. Xi, J. Lan, T. Jiang, An improved predictive controller on the FPGA by hardware matrix inversion, IEEE Trans. Ind. Electron. 65 (9) (2018) 7395–7405.

[87] M.A. Mansoori, M.R. Casu, Efficient FPGA implementation of PCA algorithm for large data using high level synthesis, in: *2019 15th Conference on Ph. D Research in Microelectronics and Electronics (PRIME)*, IEEE, 2019, pp. 65–68.

[88] D. Fernandez, C. Gonzalez, D. Mozos, S. Lopez, FPGA implementation of the principal component analysis algorithm for dimensionality reduction of hyperspectral images, J. Real-Time Image Proc. 16 (5) (2019) 1395–1406.

[89] V. Milutinović, J. Salom, N. Trifunović, R. Giorgi, Guide to Dataflow Supercomputing, Springer, 2015.

[90] C. Li, On rate region of caching problems with non-uniform file and cache sizes, IEEE Commun. Lett. 21 (2) (2016) 238–241.

[91] M.J. Flynn, O. Pell, O. Mencer, Dataflow supercomputing, in: 2012 22nd International Conference on Field Programmable Logic and Applications (FPL), IEEE, 2012, pp. 1–3.

[92] V. Milutinovic, Surviving the Design of a 200MHz RISC Microprocessor, IEEE Computer Society Press, Washington DC, USA, 1996.

[93] A. Grujic, M. Tomasevíc, V. Milutinovic, A simulation study of hardware-oriented DSM approaches, IEEE Parallel Distrib. Technol. Syst. Appl. 4 (1) (1996) 74–83.

[94] V. Milutinovic, Comparison of three suboptimum detection procedures, Electron. Lett. 16 (17) (1980) 681–683.

[95] S. Stojanović, D. Bojić, V. Milutinović, Solving Gross Pitaevskii Equation Using Dataflow Paradigm, The IPSI BgD Transactions on Internet Research, 2013, p. 17.

[96] P. Knezevic, B. Radnovic, N. Nikolic, T. Jovanovic, D. Milanov, M. Nikolic, V. Milutinovic, S. Casselman, J. Schewel, The architecture of the Obelix—an improved internet search engine, in: *Proceedings of the 33rd Annual Hawaii International Conference on System Sciences*, IEEE, 2000, p. 11.

[97] R. Trobec, R. Vasiljević, M. Tomašević, V. Milutinović, R. Beivide, M. Valero, Interconnection networks in petascale computer systems: a survey, ACM Comput. Surv. 49 (3) (2016) 1–24.

[98] O. Mencer, Maximum performance computing for exascale applications, ICSAMOS (2012).

[99] S. Sankaranarayanan, J.J. Rodrigues, V. Sugumaran, S. Kozlov, et al., Data flow and distributed deep neural network based low latency IoT-edge computation model for big data environment, Eng. Appl. Artif. Intel. 94 (2020) 103785.

[100] B. Schroeder, G.A. Gibson, Understanding failures in petascale computers, J. Phys. Conf. Ser. 78 (2007) 012022. IOP Publishing.

About the authors

Miloš Kotlar received his B.Sc. (2016) and M. Sc. (2017) degrees in Electrical and Computer Engineering from the University of Belgrade, School of Electrical Engineering, Serbia. He is a Ph.D. candidate at the School of Electrical Engineering, University of Belgrade. General research fields include implementation of machine learning algorithms using the dataflow paradigm, such as FPGA and ASIC accelerators, as well as meta learning for anomaly detection used in AutoML frameworks.

Marija Punt received her B.Sc. (2004), M. Sc. (2009), and Ph.D. (2015) degrees in Electrical and Computer Engineering from the University of Belgrade, School of Electrical Engineering, Serbia. She is currently an assistant professor at the University of Belgrade and teaches several courses on computer architecture and organization, web design, and human–computer interaction. Her research interests include computer architecture, digital systems simulation, consumer electronics, data analysis, and human–computer interaction.

Veljko Milutinović received his Ph.D. from the University of Belgrade in Serbia, spent about a decade on various faculty positions in the USA (mostly at Purdue University and more recently at the University of Indiana in Bloomington), and was a co-designer of the DARPAs pioneering GaAs RISC microprocessor on 200 MHz (about a decade before the first commercial effort onthat same speed) and was a co-designer also of the related GaAs Systolic Array (with 4096 GaAs microprocessors). Later, for almost three decades, he taught and conducted research at the University of Belgrade in Serbia, for departments of EE, MATH, BA, and PHYS/CHEM. His research is mostly in datamining algorithms and dataflow computing, with the emphasis on mapping of data analytics algorithms onto fast energy efficient architectures. Most of his research was done in cooperation with industry (Intel, Fairchild, Honeywell, Maxeler, HP,IBM, NCR, RCA, etc...)

CHAPTER FIVE

A runtime job scheduling algorithm for cluster architectures with dataflow accelerators

Nenad Korolija[a], Dragan Bojić[a], Ali R. Hurson[b], and Veljko Milutinović[c]
[a]School of Electrical Engineering, University of Belgrade, Belgrade, Serbia
[b]Department of Electrical and Computer Engineering, Missouri University of Science and Technology, Rolla, MO, United States
[c]Department of Computer Science, Indiana University in Bloomington, IN, United States

Contents

Advances in Computers, Volume 126
ISSN 0065-2458
https://doi.org/10.1016/bs.adcom.2022.01.002

Abstract

This article discusses specialized computer cluster architectures for high performance computing that include both control-flow and DataFlow components, as well as their runtime scheduling algorithms. A novel optimal scheduling algorithm for such architectures is proposed. The proposed algorithm is general, but is limited in some cases due to its time complexity. From the base optimal algorithm, two additional heuristic algorithms are derived, and then compared to other schedulers. These heuristic algorithms produce near-optimal schedules for both DataFlow hardware jobs and control-flow jobs at large job counts, with negligible scheduling penalty. Compared to an optimal scheduler, the performance gain decreases slightly as job count increases. This research illustrates that the performance of existing cluster structures can be considerably improved by adding appropriate DataFlow accelerators and a proper scheduling algorithm, while at the same time decreasing the system transistor count and power consumption.

Abbreviations

ALU   arithmetic logic unit
CFD   computational fluid dynamic
CPU   central processing unit
DAG   directed acyclic graph
DFE   dataflow engine
FCFS   first-come-first-serve
FIFO   first-in-first-out
FMem   fast memory
FPGA   field-programmable gate array
GPU   graphics processing unit
HPC   high performance computing
LBM   Lattice-Boltzmann method
LMem   large memories
PCIe   peripheral component interconnect express
PE   processing element

1. Introduction

As computer performance increases, so does the demand for executing high performance computing (HPC) applications. These applications include simulations in the domains of fluid dynamics, geophysics, biophysics, meteorology, computational finance, chemistry, etc., as well as the emerging domains of big data on the Internet of Things and Wireless

Sensor Networks. Regardless of the favored architecture, today's HPC problems are solved not by a single node computer, but by clustering computers together.

Cluster computing is a form of computing in which independent computer systems (often called nodes) are connected through a local area network so that they behave like a single machine. In most circumstances, all cluster nodes use the same type of hardware, which usually consists of central processors and auxiliary hardware accelerators. Each node performs a specific job, controlled, and scheduled by software. By splitting an application execution in multiple jobs over many processing nodes, a computer cluster can solve complex operations more efficiently and with better data integrity, than a single computer.

Arranging the execution of jobs on a cluster requires a scheduling algorithm. A common trait for most of the existing HPC schedulers is that they can schedule an instantiation of a single application on each node of a cluster, or a single application at any given time in the case of reconfigurable computing. Often appropriate resources for an application are planned by the system designer, so that there is no need for a runtime scheduler. Scheduling a mix of heterogeneous applications in a computing cluster with special hardware accelerators is a relatively recent problem. In the case of DataFlow programming, some of the applications are designed for execution on control-flow hardware, where others might be executed both on control-flow and on hardware accelerators. The primary goal of a scheduler is to minimize the total execution time of all jobs. However, the scheduler requires processing time and computing resources to determine a schedule. Optimal job scheduling is known to be NP hard computational problem. As a result, heuristics are often designed to create suboptimal schedules, while consuming less processing power.

In this article, we present an optimal scheduling algorithm for cluster computer architectures with nodes that combine control-flow and DataFlow hardware. Since the problem is NP hard, two heuristic algorithms are derived that produce near-optimal schedules for an unlimited number of jobs, with a negligible scheduling penalty.

The introduction further discusses DataFlow systems, their architectures and programming languages. Next, the problem of scheduling is defined, and various related scheduling techniques are discussed. Following the problem definition, new scheduling algorithms are presented. Finally, scheduling details for optimal and near-optimal scheduling algorithms are articulated, followed by simulation results and relevant discussions.

1.1 DataFlow architectures

The DataFlow computation paradigm deviates from the conventional control-flow paradigm in two fundamental ways: asynchrony and functionality [1,2]. While control-flow instructions are executed sequentially under the control of a program, any DataFlow instruction can be executed when all the required operands are available. In DataFlow, any two enabled instructions that do not interfere with each other can be executed concurrently.

A DataFlow graph, a kind of directed graph, where nodes represent instructions and arcs represent data dependencies among the nodes, can be viewed as the machine language for a DataFlow computer, as is the case with the compiled program for control-flow computer [2]. The basic abstract elements of a DataFlow graph, an operator, a decider, a copy actor, a merge actor, and a switch actor, are shown in Fig. 1. These elements correspond to digital logic elements that enable data flow through the hardware:

(1) basic logic elements (AND, OR, NOT, XOR, etc.) form logical functions, i.e., arithmetic logic units (composed of logic elements) responsible for machine instruction execution

(2) multiplexers select an input that should be directed to an output

(3) demultiplexers select an output where an input should be directed.

In a DataFlow graph, the operands are conveyed from one node to another in data packets, often called tokens. A data value produced by an operator is directed by means of either a merge or a switch actor. A merge actor passes an input token determined by the control signal to the output. A switch actor passes an input data token to the output line determined by the control signal.

An abstract DataFlow model assumes that tokens carry data values along arcs. In the ideal case, arcs are infinite first-in-first-out (FIFO) queues for

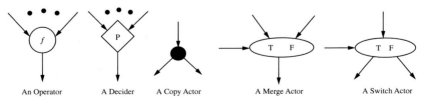

An Operator   A Decider   A Copy Actor   A Merge Actor   A Switch Actor

Fig. 1 Basic elements of a DataFlow graph.

keeping tokens with data until they are required for further processing. Practically, these queues have limited capacities. Based on the size of the queue, two main implementations in early DataFlow systems were identified, static and dynamic DataFlow architecture. Static DataFlow architectures define arcs as single token containers. On the other hand, dynamic DataFlow architectures define arcs as token queues. In either case, an instruction is enabled for execution if all required input tokens are ready and there is an empty slot for the output token. The instruction execution generates output tokens, but also clears its input tokens.

Static architectures limit the number of instructions that can be executed in an array of operators connected by arcs. This limits performance, especially in the case of sequential code execution within loops. One form of the static DataFlow architecture is represented by the Static DataFlow machine developed at MIT [3–5]. Dynamic DataFlow architecture forms are represented by Tagged-Token DataFlow architecture and Manchester machine [3–5]. These implementations enable more parallelism to be exploited and out of order execution, avoiding of stalls in parts of the architecture which wait for tokens that are computed slower comparing to most of others. However, implementations require more complex logic and associative memory for matching tokens.

Early DataFlow systems have advantages over the more mature control-flow systems which are simpler and more suitable for expressing parallelism and data dependencies [6]. DataFlow systems support two types of parallelism: spatial and pipeline. Spatial parallelism allows any two nodes to execute instructions simultaneously if there are no data dependences between them. Pipeline parallelism allows separate calculations to be performed at different stages in a pipeline. However, DataFlow systems have drawbacks, which is why early DataFlow systems were not widely accepted. There is a large overhead in instruction execution and managing storage resources for data structures. Also, most applications do not have enough parallelism to exploit DataFlow potentials appropriately. Hardware processing resources were often wasted, resulting in a lower fraction of executed floating-point instructions in numerical programs, compared to control-flow multiprocessor systems, which benefited from decades of emerging advances [7–12]. In addition, DataFlow systems are very ineffective for sequential code execution. As a result, hybrid approaches have been proposed, including both control-flow processors and the support for the DataFlow model [6], aided by compilers and adequate program representations [13].

Maxeler's multiscale DataFlow computing systems is a modern representative of the DataFlow concept implemented using FPGA technology [14].
They are built around the DataFlow engines (DFEs) that consist of the
following elements: FPGA chips, which are comprised of thousands of
logical components, fast on-chip memories (FMems) that can deliver data
in one clock cycle with capacities of several Mbs, and large off-chip memories (LMems) up to 48 GBs but slower than FMems. The DFEs also have
logic for interconnecting with the central processing unit (CPU) through
a Peripheral Component Interconnect Express (PCIe) interface or with
other DataFlow engines over a MaxRing connection. FPGA technology
is much slower than the technology used in traditional control-flow processors. However, it uses much less power and space, and is therefore able to
accommodate packing more smaller devices together due to reduced
cooling requirements [15]. Fig. 2 shows the architecture of a Maxeler
DataFlow system.

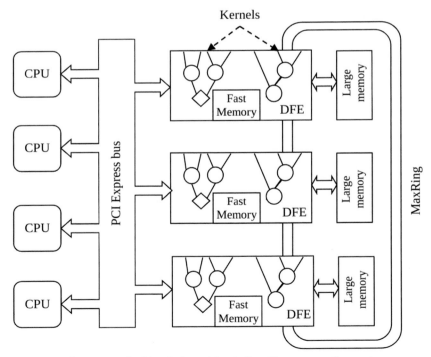

Fig. 2 An architecture of a Maxeler's multiscale DataFlow computing system.

The internal organization of the DataFlow engine can include multiple kernels, which consist of DataFlow execution graphs. In every cycle data move through DataFlow elements which execute simple arithmetic and logic operations and propagate results to other elements according to the DataFlow graph. For a static DataFlow concept, data dependencies are resolved at compile time. This architecture inherently exploits parallelism extensively, enabling spatial computation and parallelism within a deep pipeline structure, allowing multiplying kernels to overlap computation, and queued arcs to overlap communication with the CPU host code execution or DataFlow kernels execution. Also, data can be preloaded to either FMem or LMem for reuse during calculations, thereby avoiding waits when fetching data.

1.2 DataFlow languages

DataFlow computing requires the design of special languages, so that the hardware characteristics of DataFlow machines can be addressed appropriately. These special languages must support control-flow computer architecture programming as well so that the initialization phase of a scalable algorithm could be executed on control-flow architecture, while the DataFlow hardware could be used for parallel parts of the algorithm. Meanwhile, the basics of a DataFlow language should permit the specification of algorithms that can exploit DataFlow principles [1].

These principles should ensure fixed scheduling of data represented by arcs in the DataFlow graph, as determined from data dependencies and variables that cannot have different values once the value has been assigned. To ease the transition for programmers who are typically accustomed to the control-flow way of thinking, DataFlow languages resemble traditional functional languages with constructs such as loops that have imperative syntax [16]. Typical representatives are Value Algorithmic Language, Irvine DataFlow Language, Stream and iteration in a single-assignment language, LAPSE, and LAU [5,13]. Since a DataFlow graph is a machine language of DataFlow systems, it is natural to describe the DataFlow algorithm graphically. As a result, since the 1990s the development of visual DataFlow programming languages has come into focus in the research community [13].

An application for Maxeler's multiscale DataFlow computing systems must contain programs for the DataFlow engine (kernels) and for managing streams between the CPU host and a DFE, and among multiple DFEs.

Maxeler Technologies has developed a specialized language, MaxJ, which has a subset of Java programming language functionality and the support for programming DFE kernels and connecting them to the CPU [17]. Algorithm 1 below presents an example kernel used for calculating an average for each tree of successive elements in an array. As it can be seen, even ternary operators are supported by the MaxJ. A limitation of the language, or the hardware itself, is that a hardware variable could be assigned the value of a program variable, but not vice versa. In addition, once the value is stored into the hardware variable, it cannot be restored to the program variable. The reason for this is easy to comprehend. Values are translated into set voltages, but as signals flow through the hardware the voltages change, hence values cannot be translated back to program variables.

There are many optimization techniques that can be applied to algorithms developed for standard CPU and graphics processing unit (GPU) computing systems. But for DataFlow, developers should also exploit the parallelism available in deep pipelining the DFE graph, extensively utilizing spatial computation, overlapping computation with communication, reorganizing code execution to avoid stalls in the execution, storing data in FMem or LMem for later reuse, and reducing the fixed-point size of the data stream format and increasing the number of iterations if necessary, to achieve the same or better accuracy.

2. Problem statement: Scheduling jobs on clusters with DataFlow hardware

Scheduling is the operation of assigning jobs for execution on hardware resources such as processors, or DataFlow accelerator cards. The jobs may be virtual computation elements such as threads, processes or DataFlow kernels. While control-flow jobs duration is hard to predict, DataFlow jobs often consist of a number of iterations known prior to execution, where iteration duration can be calculated based on the DataFlow pipeline depth (the time needed for executing longest DAG path of one iteration in the kernel). Further, it is possible to estimate execution time of a control-flow job knowing the amount of processing and the speed of the hardware. The focus of our research is scheduling for cluster architectures control-flow jobs and DataFlow jobs with estimated execution time.

A scheduler can have multiple goals, for example: maximizing the throughput (the total amount of work completed per time unit); minimizing wait time (for work to become ready before execution); minimizing latency (response time); or including deadlines in real-time systems.

Most computer clusters do immediate dispatching (also known as task assignment), whereby incoming jobs are immediately assigned to servers (also known as nodes). Such clusters require the task assignment policy in the form of rules that are used by the front-end router (also known as a dispatcher or load balancer) to assign incoming jobs to servers. For example, incoming jobs may be assigned to servers in round-robin order, or each incoming job might be assigned to the server with the shortest queue [18].

The scheduling policy deployed at an individual host is not fixed but is typically application dependent. First in first out (FIFO), also known as first come, first served (FCFS), is the simplest scheduling algorithm. FIFO queues processes in the order that they arrive in a queue. Other strategies include Round Robin, Shortest Remaining Time First, Fixed Priority Preemptive, etc.

Scheduling methods can be partitioned into:
- Static scheduling, or compile-time scheduling and
- Dynamic scheduling, or runtime scheduling.

A scheduling algorithm is static if the scheduling decision (what computational tasks will be allocated to what processors) is made before runtime. A scheduling algorithm is dynamic if the decision is made at runtime. Static scheduling is used in application-specific computing for equipment that should continuously execute the same set of operations on data relatively fast, e.g., network-on-chip, where data is streamed through hardware that serves only a single purpose. Although relatively fast, lack of flexibility is its drawback.

Clusters with reconfigurable DataFlow elements require dynamic scheduling for reconfiguring FPGAs at runtime. Unfortunately, this imposes additional processing that presents a runtime overhead. Various algorithms use heuristics to make scheduling times acceptable, by producing schedules that are sub-optimal.

Scheduling in clusters is inherently dynamic and some clusters are composed of control-flow and DataFlow hardware. Some researchers have tried replacing CPUs by smaller processing elements (PEs) to efficiently schedule executions and memory accesses [19]. The process of applying a new resource (e.g. dataflow paradigm) to the old problem of scheduling execution on processing elements could be considered as creativity based on implantation [20].

Next is a brief survey of various cluster scheduling techniques. The survey begins with static and dynamic scheduling techniques for multiprocessor architectures with either homogeneous (control flow based) or heterogeneous nodes. It follows with techniques that specifically relate to DataFlow based cluster architectures, implemented with reconfigurable elements (FPGAs).

2.1 Static multiprocessor scheduling

In static scheduling, a parallel program can be represented by a directed acyclic graph (DAG), also known as the coarse level DataFlow graph, consisting of a set of nodes that represent task execution, and a set of directed edges that represent the communication among tasks. The direction of edges captures the precedence constraints among the tasks. Optionally, the graph could be weighted, where a node weight denotes the computation cost of the task and an edge weight denotes the communication cost. The source node of an edge can be treated as a parent node, and the destination node can be treated as a child node. A root node, having no parent node, represents the beginning of the algorithm, or the entry node, and a node that has no child represents the end of the algorithm execution, or the exit node [21].

The target platform is a distributed memory multiprocessor system where each processor has its own local memory. Processors communicate with each other through explicit messages among them. The communication cost is only considered for data token messaging between processors. The objective of scheduling is to minimize the total execution time of all processors by assigning tasks to processors so that the precedence constraints are preserved.

Scheduling algorithms can be divided into the following four categories: (1) *Bounded number of processors*, scheduling a DAG to a limited number of processors; All the processors are fully connected and hence the communication cost is negligible. (2) *Unbounded number of processors*, scheduling a DAG to an unbounded number of processors; These processors may be mapped onto multiple clusters using a separate mapping algorithm that assumes the processors within clusters are fully connected. (3) *Arbitrary processor network*; A scheduler explicitly schedules communication messages on the network channels to avoid network congestion. (4) *Task duplication-based scheduling* replicates tasks to reduce the communication overhead. It should be noted that duplication can also be applied in any other scheduling categories.

2.2 Dynamic multiprocessor scheduling

In contrast to static scheduling, the dynamic scheduling is applied when some or all the task characteristics are not known a priori. For example, for event-driven applications the scenario is not known during compile time, because it is impossible to predict events before execution. The tasks are characterized by arbitrary arrival times. Another example is numerical applications in which the input to the application consists of largely varying data [22].

Ramamirtham et al. proposed the Myopic algorithm [23]. At any time, a task can arrive at any node in the system. The node scheduler tries to schedule the task to be completed before its deadline, along with other previously guaranteed tasks on the node. The new task will be guaranteed, if and only if a schedule that fulfils all deadlines can be determined. A task that cannot be guaranteed can be sent to another node to find a schedule that guarantees execution before its deadline. In either case, previously guaranteed tasks remain guaranteed.

Lee et al. created a scheduling algorithm targeted for DataFlow multiprocessors based on a *vertical layering method* for the DAG [24]. The DAG is first partitioned into a set of vertical layers of nodes. The initial set of vertical layers is built around the critical path of the DAG. A vertical layer consists of data dependent nodes. Later, in the optimization phase, various cases are considered, to find the DAG where inter-processor communication on a portion of DAG would not be longer lasting than the computation in a single layer. Finally, the vertical layers of nodes are mapped to processors that minimize the schedule length.

The *self-adjusting dynamic scheduling algorithm*, proposed by Hamidzadeh et al. [25], uses a unified cost model to account for issues such as processor load, memory locality, and scheduling overhead at runtime. A separate processor is responsible for scheduling, and it maintains a tree of partial schedules and incrementally assigns tasks to the least-cost schedule. Scheduling terminates whenever any processor becomes idle. At this moment, the partial schedules must be distributed to the processors. This scheduling algorithm performs well for a small number of processors. In addition, it does not scale well when the mean time of task executions is small.

In the *minimal earliest finish time algorithm* for heterogeneous multiprocessor systems [26], a real-time task is assigned to the processor on which its "earliest finish time" is minimal. Once scheduled, the task cannot migrate to another processor.

2.3 Scheduling for heterogeneous CPU/FPGA architectures

Abdessamad et al. [27] model the scheduling problem mathematically in a heterogeneous computing architecture that includes FPGAs. The goal is to achieve real-time capabilities and high performance for simulations in the avionic industry. They present a solver based on the Mixed Integer Program for scheduling based on the assumption that task graph topology, execution time, and communication delays are known before scheduling a task. Their notion of a task corresponds to our notion of job, which is a compute kernel prepared to run on an FPGA. The algorithm can schedule up to 50 tasks within a few seconds. A drawback is that it does not necessarily find the best schedule. Since DataFlow hardware tasks are often relatively long-lasting, it may be reasonable to employ a slower scheduler that creates more effective schedules.

Hamilton et al. [28] presented a multitasking runtime system on a CPU/FPGA platform, where user applications are executed as mixed-architecture processes. Their underlying platform also supports context switching mechanisms, blocking, and preemption. The goal was to support concurrent execution of mixed-architecture processes. This algorithm is not specifically created for high performance computing and would require modification to produce better schedules for HPC, without allowing pre-emption for FPGA images, where any DFE image holds hardware imple-mentation details of task that can run using a single DataFlow hardware element. Preemption between consecutive FPGA image runs could be allowed.

Khuat et al. [29] introduce a runtime spatial-temporal scheduling for dependent tasks (i.e., jobs) prefetching on heterogeneous FPGA-based systems to reduce the reconfiguration delay of tasks. Depending on the availability of DataFlow hardware, this approach could be very useful in high performance computing. While CPUs are executing their tasks, DataFlow hardware can be prepared for executing jobs that are most likely to occur in the near future.

Cilardo et al. [30] proposed a methodology for optimization of FPGA-based, many-core interconnect architectures. Although this approach does not produce a schedule, it can be useful for connecting DataFlow hardware with CPUs.

Iturbe et al. [31] considered scheduling real-time tasks (i.e., jobs) on FPGAs. They presented eight simple algorithms, suitable for a relatively high number of tasks, each having relatively short durations. Our concern is that more complex processing is needed for better scheduling at a cost of even more scheduling time.

Fei et al. proposed a scheduler based on solving a task shop scheduling problem [32]. Their intelligent optimization algorithm is implemented using FPGA.

2.4 Dynamic scheduling for heterogeneous CPU/FPGA architectures

There are two approaches widely used to solve scheduling problems with durational uncertainty that can be found in the literature. One computes an initial baseline schedule off-line and modify it (if required) during the execution of jobs, as demonstrated by Van de Vonder et al. [33]. The second approach creates policies to provide on-line decision rules for determining which jobs to start at the moment of scheduling, and which resources to assign [34]. Scheduling jobs at compile time is usually not possible due to the unknown durations of jobs, as well as unknown starting times, so the on-line approach is more appropriate for the targeted applications in this article. Before the scheduler is started, one can determine which jobs fit together into an FPGA image (multiple compute kernels loaded together onto the FPGA). Then at runtime, depending on the size of data each job processes, the scheduler can select the best suited FPGA images, and can schedule the execution of jobs on DataFlow hardware, accordingly. This approach is useful for another reasons, as well. Since DataFlow hardware might not be the bottleneck of the system, it could be configured for producing schedules in the same way that CPUs produce schedules. Hence, depending on the bottleneck, either CPUs or DataFlow hardware can be used for creating schedules.

Yu et al. [35] introduced a novel approach for scheduling tasks on networked multi-FPGA systems at the runtime. The scheduler maintains high performance even for irregular tasks, i.e., it scales well even in the case of significant granularity of the tasks and/or if the number of computational units is relatively large. The scheduler works on the principle of topological ranking, allowing realistic irregular workloads to be processed with a significantly higher level of performance compared to the existing schedulers.

3. Essence of the proposed solution and its potentials

This section will describe two new branch-and-bound based heuristic schedulers that schedule jobs onto DFE images using recursion, while bounds are exploited to prune the searching space. The aim of the first one is to minimize the total execution time of all applications. The second one maximizes throughput. Each of these two schedulers can be used as a part of the cluster scheduler. After explaining the basic idea, we will define the analytical model and discuss issues of importance. Finally, scheduler implementation details will be presented.

We are targeting DataFlow computing models with a set of interconnected nodes that process data. In such a reconfigurable computer architecture, hardware jobs are placed on a set of DFEs, where each DFE represents a reconfigurable chip with a relatively high amount of memory (see Algorithm 1). The CPU is responsible for sending a job specification, and the scheduler determines whether the job should be executed on the CPU or DFEs.

The most important difference between previously mentioned single node DFE scheduling and traditional scheduling for reconfigurable clusters is in the amount of resources planned for a single job. Traditional scheduling for reconfigurable clusters assigns, in the general case, multiple nodes to a single job. If a job requires only relatively small amount of certain resource available on each DFE, the DFEs are left underutilized, i.e., specific areas of FPGAs are not assigned to any job that is available for scheduling. Our goal is to investigate how to share single reconfigurable computing unit as a resource that will be utilized by multiple processors for running jobs that predominantly require different types of resources available on DFEs.

In DataFlow programming, applications are typically developed based on the already existing control-flow application. If both types are available in a DataFlow cluster, a scheduler can decide whether the application will be executed on a CPU, or on a DFE. The best schedule for a single DFE is found by comparing the total CPU and DFE computation time for all possible scenarios, i.e., when each subset of jobs that can run simultaneously on a single DFE image is scheduled for a DFE, while other jobs are scheduled for a CPU.

Based on recursive schedulers for finding optimal solutions in terms of the total execution time and throughput, we have defined and implemented schedulers based on heuristic scoring capable of scheduling an arbitrary number of jobs. For each job that can be scheduled on both the CPU and the DFE, we will assign a score. A score depends on:

- time needed for the DataFlow job to be processed by the CPU,
- time needed for the DFE to process it,
- time needed for initializing the job to be executed on the DFE,
- the amount of resources to be consumed for executing all jobs using the CPU, and
- the amount of resources to be consumed for executing all jobs using DFE.

Since DFE can be configured to execute many jobs in parallel and the time needed for each job to be processed differs, a score assignment is made at runtime.

Our cluster scheduler assigns all jobs to nodes. Nodes schedule jobs according to the heuristic score-based scheduling. Splitting is done one job at a time, based on resources that a job requires and availability of resources on nodes. Schedulers assign a certain number of DFEs for data processing.

3.1 Analytical modeling of the proposed solution for scheduling jobs

It is a challenge to efficiently determine which jobs should be placed onto DFEs to minimize the total execution time of all jobs. The following is a formal definition of the job scheduling problem on both DataFlow hardware and CPU. The set of DataFlow jobs available for scheduling at a given moment t is defined as:

$Jt = \{Jt\_1, Jt\_2, \ldots Jt\_nJobs\}$, where $nJobs$ is the number of jobs available at time t.

Each job Ji is defined by:

TCi—*estimated time needed for CPU to process job i,*

TDi—*estimated time needed for a DFE to process job i,*

TIi—*estimated time needed for initializing job i to be executed on the DFE,*

$TCMi$—*the amount of memory that job i needs for execution on the CPU,*

$TDMi$—*the amount of memory that job i needs for execution on the DFE, and*

$TDDi$—*the amount of computing resources that job i needs, to be executed using DFE.*

The best schedule is defined as:

$B = \{Jtb\_1, Jtb\_2, \ldots Jtb\_m\}$, where at time t,

Jtb_i represents the i-th job in the best schedule containing m jobs. Note that beside TDi, TCi can also be estimated based on the amount of processing known in the runtime and the speed of the hardware.

Each DFE can be used to execute only a limited number of jobs. Any DFE image holds hardware implementation details of a collection of jobs that can run simultaneously on a single DataFlow hardware element. A scheduler may produce an arbitrary number of consecutive DFE images. Each DFE image should stay loaded in the hardware until all of its jobs are processed. Let us define the set of jobs included in a DFE image i as:

$Ii = \{JIi\_1, JIi\_2, \ldots JIi\_k\}$, where k is the number of jobs scheduled for the DFE image Ii.

The time needed for a CPU to execute all jobs that are scheduled for execution using the CPU is equal to:

$$TCPU = \sum TCi.$$

The time needed for the DataFlow hardware to execute a DFE image i is equal to the time needed for the longest DataFlow job included in a DFE image i:

$TIi = max(TDIi1, TDIi2, \ldots, TDIik)$, where function max returns the maximum value, and $TDIim$ represents the time needed for executing the m-th job in the DFE image i.

The time needed for the DataFlow hardware to execute all scheduled jobs using the DataFlow hardware is equal to the sum of the time each DFE image needs to execute jobs it contains:

$$TDFE = \sum TIi.$$

The best schedule is defined as the schedule with the minimal value of $max(TCPU, TDFE)$. This problem cannot be solved in polynomial time and scheduling the execution of jobs must be performed at runtime. The criteria is relaxed by using the heuristic approach for minimizing the value of $max(TCPU, TDFE)$. An additional constraint is that the scheduling algorithm execution time, or part of it, should be added to the total execution time, unless the scheduler works on the CPU that is otherwise idle.

Input to the scheduler is a set of jobs to be scheduled for execution either on a CPU or a DFE. Each job that can be executed on both DFE and the CPU is defined with the following attributes:

- *%DFE*: Percentage of DFE needed to execute a job using a DataFlow hardware
- *tDFE*: Time needed for a DataFlow hardware to execute the job
- *tCPU*: Time needed for the CPU to execute the job
- *tInitDFE*: Time needed to initialize the DataFlow hardware with an appropriate VHDL file
- *memDFE*: The DataFlow hardware memory needed
- *memCPU*: RAM memory needed for the job executed on the CPU
- *memRest*: RAM memory needed for the job executed on the DataFlow hardware.

Apart from these, the additional attribute *tRest* defines the time needed for CPU to initialize an application that should be executed using the DataFlow hardware, as well as the time to collect result. High performance algorithms from the open literature used for creating synthetic benchmark for this research can be accelerated using DataFlow hardware and a CPU is required only for initialization and collecting results. Initialization duration (*Tli*) is negligible in relation to *tDFE* and we will exclude this parameter from further discussions.

3.2 Issues of importance for scheduling methodologies on reconfigurable hardware

All applications can be divided based on suitability for execution on DataFlow hardware. Applications or parts of applications (jobs) suitable for execution on a DataFlow hardware may be accelerated using DataFlow hardware architectures with acceleration factors even greater than 10 [36–39]. Many other algorithms can be accelerated using the DataFlow paradigm. All of these applications have components that must be executed on a CPU, and other components that can be accelerated using DataFlow hardware. We will refer to an application or part of an application that must be executed on a CPU as a CPU job. We will refer to an application or a part of application that can be scheduled for execution either on a CPU or a DataFlow hardware as a job. Jobs suitable for execution on a DataFlow hardware and only those jobs can run both on a CPU and on a DataFlow hardware. We will also refer to these as streaming jobs, if they can execute instructions while streaming data to and from the CPU. Data can be streamed over a PCIe bus or any type of network.

Although executing certain types of algorithms on specialized hardware is faster, configuring FPGAs during application execution adds overhead in the time domain. In the case of applications that can be accelerated using DataFlow hardware, image files can be prepared either at compile-time or at runtime. We will assume that images are prepared at compile-time, as is the case with compiled programs for control-flow processors. This way, an application written in a high-level programming language that is not suitable for DataFlow hardware can run as a job on a CPU, while other jobs can run either on the DataFlow hardware (depending upon availability), or on the CPU. The DataFlow hardware can be shared between multiple jobs at the same time only if appropriate image file is prepared for a given sets of jobs. The creation of image files can be automated, but the automation of translating multiple DataFlow applications to a single image file is out of the scope of this research. Sharing DataFlow hardware resources between multiple applications in a cluster can be done at the node level.

3.3 Details of the proposed solution for the scheduling problem

We have developed two recursive algorithms for finding the best schedule with principles similar to the ones proposed in [40]. The first algorithm aims at minimizing the total execution time of all applications. The second one aims at increasing throughput as much as possible. They are both based on the same principles, and they differ only in their scoring methods. Hence, when explaining the main principles, we will be referring to both as "the" scheduler. Starting with the empty set of jobs included in the potential schedule, the scheduler tries to incorporate each DFE job in the set. For each potential schedule, a score is assigned, based on the ratio between the total execution time of all jobs on the CPU and for running them as scheduled. For each potential schedule, the same procedure is repeated, adding new jobs to the potential schedule, evaluating scores, etc. The schedule with the highest score is considered as the best schedule. Note that potential schedule can include more than one DFE image. The score for a potential schedule of our first scheduler is calculated as the reduction in execution time on CPU alone and using both CPU and DFE, where jobs are optimally placed in an arbitrary number of DFE images, so that the total reduction in execution time is maximal. For our second scheduler, the total throughput should be maximized. The throughput is defined as the total execution time of all jobs scheduled for the DataFlow hardware when running them on the CPU divided by the total execution time of all DFE images. These two algorithms are not useful in scheduling dozens of jobs in runtime, because only a limited number of jobs can be scheduled before the scheduling time becomes unreasonably long. However, they are useful for evaluating scoring schedulers. Another metric that will be used for evaluating schedulers is the comparison of the time needed for the CPU to process all jobs scheduled to be executed on the DataFlow hardware.

The main limitation of a recursive scheduler is the number of jobs that can be considered for placing on the DataFlow hardware. Therefore, we have modified the scheduler so that it selects only k jobs from n available jobs, where $k \leq n$, as shown in Algorithm 2. After m jobs have been executed, where $m \leq k$, and new jobs became ready for scheduling, our scheduler is called again to schedule k jobs. Numbers k and m are determined experimentally, by varying their values to maximize the acceleration. The time needed for scheduling is also taken into account. Note that for $k = m$ the algorithm returns the best schedule. Although there are iterative algorithms for finding all k subsets of n elements, the recursive one best suite our need to eliminate branches that cannot produce better results than the one already found. There is another important modification we made to the best scheduling algorithm. For each job that can be added to the potential DFE schedule, two different scenarios are considered. The first one adds the job to the currently observed DFE image. Of course, the total percentage of each required resource by the considered job, together with jobs from the observed DFE image, is lower or equal to 100%. Resources include the DFE, the PCIe bus, the main memory and the memory on the DataFlow card. The simulator is extendable, so that new types of resources could be taken into the account. If the job can be added to the currently observed DFE image without affecting the execution time of the DFE image, the adding the job to the second DFE image is not considered. Function DFE returns the percentage of DFE (*%DFE*) that a job or the set of jobs occupy. Similarly, the *PCIe* function returns the percentage of PCIe (*%PCIe*). Function *mem* determines if there is enough main memory and DataFlow card memory for a job or a set of jobs.

The score functions in our heuristic algorithms are modified so that the reduction in time needed for jobs included in one DFE image is multiplied by a scaling factor. The scaling factor is further divided by 1.1, an empirically derived number for each DFE image scheduled for later execution. This assures that the total reduction in the execution time of a first DFE image is valued more than the same reduction in the execution time of the following DFE images. The reason behind is that, as time passes, it becomes more likely that new jobs may change the schedule for any DFE image scheduled for later run.

The scheduler initially assumes no jobs are included in the schedule. For a given depth k, the scheduler recursively searches for all possible combinations of jobs in the schedule, associating a score to each of them. The one with the highest score is treated as the best possible solution. The first m jobs from the schedule (if available) are scheduled for the execution. After a given number of jobs are executed, rescheduling is performed. With this rescheduling, jobs that were included in the schedule might not exist in the next schedule.

Consider the scenario where most of the jobs can be executed using a DataFlow hardware. In this case the scheduler can pick any k jobs from a relatively high number of jobs n, leading to relatively long scheduling time. Therefore, each job is assigned an initial score. This score represents how effective it would be to schedule the job on the DataFlow hardware without taking into account jobs already assigned to the DataFlow hardware. If certain jobs can be accelerated significantly more than others using the DataFlow hardware, there is a relatively high probability that they should be placed on the DataFlow hardware. The following modification to the starting recursive scheduler is that jobs are sorted first according to their initial score. The scheduler will consider putting jobs with highest scores on the DataFlow hardware. This way, including jobs that are not likely to produce the best schedule will halt the recursion before the depth of k. The initial score is assigned the value derived from the reduction in the execution time of a job that uses the DataFlow hardware and the execution time of the same job using the CPU. Then, this value is divided by the factor derived from the amount of resources it would consume on the DataFlow hardware, and then multiplied by the factor for the difference in the amount of resources the job would consume on the CPU while running on the DataFlow hardware, and for running it solely on the CPU.

Previous modifications can even be used for scheduling jobs during compile time and storing the best scenarios for each combination of jobs in a hash table. Each hash table entry can be found as a function of sets of jobs that were available for scheduling. Although this solution can be relatively fast in some cases, this approach is not always useful, especially if the number of possible combinations is too large to keep all potential DFE images. For example, if a CPU has only a small amount of available RAM memory, jobs that would consume too much memory if executed on the CPU should be rather placed on the DataFlow hardware. Accordingly, the initial score is modified during runtime. Depending on the bottleneck of the system at the moment of scheduling jobs, initial scores of jobs that require fewer resources and present a bottleneck should be increased.

If two jobs with different duration are to be scheduled on the DataFlow hardware, even if they can both fit and share resources without affecting their performance, it might be better to schedule the first one separately and combine the second one with some other job available for scheduling. Hence, it is important not only to choose a subset of k jobs for scheduling, but also to determine which jobs should be packed and executed together.

However, if a job can fit within the DataFlow hardware along with other jobs, and its duration is less than the longest executing job prepared for that image, it is immediately packed together with other jobs. This reduces the amount of computation needed for scheduling.

The main part of the scheduler is implemented in the C++ and is shown in Algorithm 3. This algorithm finds the best schedule for the DataFlow hardware, without taking into consideration the number of jobs that will have to be scheduled on the CPU.

Our cluster scheduler design is based on the above scheduling approaches. The cluster scheduler aims at splitting all jobs available for scheduling in the cluster onto the nodes, where each node uses the previously described scheduler. Splitting is done one job at a time, based on resources that a job requires. In order to find an appropriate node where the job should be sent, it is necessary to keep track of the average usage of each resource. Assume that resources in a cluster are uniformly occupied. If a job requires 20% of the first resource of a node, and 40% of the second resource, nodes with higher availability of the second resource would be considered first. This approach unifies resource availability. If jobs on average require more resources of the one type, then the cluster scheduler would first consider nodes with lower occupation of that resource.

In the following chapter scheduling simulations for various scheduling scenarios are performed using parameters that affect scheduling on a single node. Results are analyzed and discussed.

4. Experimental analysis of the proposed solution

Representative algorithms from HPC applications used for benchmarking are explained here. All selected algorithms are implemented using the Maxeler framework and are briefly discussed. Scheduling details are explained. The results of scheduling jobs from the synthetic benchmark are presented, followed by a discussion of single-node and cluster-scheduling. Finally, the validity of the results is discussed.

4.1 Benchmarks and datasets used in this analysis

To evaluate scheduling acceleration of applications using a CPU, DataFlow hardware, and an appropriate scheduler, we have included a comparison of execution times of various algorithms on both the CPU and the Maxeler DataFlow hardware.

There has been research on a wide range of algorithms using the DataFlow approach. Here, we focus on those works that include the implementation of algorithms using the Maxeler framework. The proposed schedulers will be tested against a combination of applications that can run only on CPUs and those that achieve acceleration using the Maxeler framework.

Computational fluid dynamics (CFD) methods [41] are used for simulating fluid flows. Traditional methods are based on solving the Navier–Stokes equations. The Lattice-Boltzmann method (LBM) models the fluid as a mesh consisting of particles that perform consecutive propagation and collision processes repeatedly [42]. Calculating influences on each particle in the volume the simulation time requires a significant amount of computation. The LBM method is designed to be scalable and capable of running efficiently on massive parallel computer architectures. In our previous work [36], the Bhatnagar-Gross-Krook model (using the single relaxation time approximation) [43] is used for simulating blood flow. We have achieved acceleration factors of 15–100, over a range of problems. However, we did not account for the time needed to prepare the DataFlow hardware for executing the algorithm. Hence, the high-end acceleration factors (~ 100) were achieved only for a very small number of iterations and can be explained by the slower processing time of the CPU due to data fetches from main memory. Since the time needed for preparing the DataFlow hardware is a few orders of magnitude less than the algorithm execution time needed for calculating blood flow, we can consider the acceleration factor to be 15–17, based on acceleration graph given in fig. 26 of the paper [36].

The Gross-Pitaevskii Equation is a widely used model in the context of Bose-Einstein Condensates. It is based on local interactions between particles. Stojanovic et al. solved numerically the Gross-Pitaevskii Equation using the Maxeler DataFlow hardware and achieved the acceleration factor of 8.2 [19]. Unlike the control-flow implementation of the algorithm, the DataFlow implementation needed rearrangement of input data to exploit the DataFlow hardware, by passing input data in consecutive cycles. For big problems, instead of passing elements of a single row that are all dependent on the previous elements, they pass only the first elements of many independent rows, and then only second elements of such rows, etc. In this way, all of the elements have the data ready once the processing is scheduled.

The Odd-Even Merge Network Sort algorithm is a sorting algorithm that includes a relatively high number of two-number comparisons that can be done in parallel. It was designed by Ken Batcher for sorting networks of size $O(n*(log\ n)2)$, but it gained popularity with GPU processing. Kos et al. implemented the Odd–Even Merge Network Sort algorithm using the Maxeler DataFlow hardware, and achieved acceleration factors of 100–150, when sorting more than 1000 arrays consecutively [37]. Each of the arrays consisted of 16–128 elements. Their implementation of the Odd-Even Merge Sort algorithm in parallel sorts the elements of an array in 10–28 steps.

A spherical code is a set of n-dimensional vectors on the unit sphere, whose coordinates are real numbers. The associated packing problem is regarded by researchers as one of two standard optimization problems associated with spherical codes [44]. Given a set of vectors and a number of dimensions, the goal is to find a spherical code (vectors) that has maximal value for the minimum Euclidean distance between two code vectors. They achieved acceleration factors between 18 and 24, by implementing in the Maxeler DataFlow hardware the control-flow implementation of the Spherical Code Design, based on the variable repulsion force method. Although they have changed the original algorithm, the iterative approach remains.

The RSA public-key cryptosystem is based on the product of two large prime numbers, and the fact that there is no known algorithm that can efficiently find prime factors in the product. Public key is spread for encryption, while the decryption key is private and used for the decryption process. For example, a entity can give its public key to users, so that they can encrypt their data and send the data back to the entity that is the only one able to decrypt it. Bezanic et al. [39] implemented the RSA algorithm using the Maxeler DataFlow hardware, achieving the speedup factor of 25%–30% for encrypting data bigger than 40 MB. Although DataFlow hardware has good potentials, the acceleration factor is limited in this case, since only a part of the execution time that is spent on those instructions can be executed efficiently using the DataFlow hardware.

Algorithms and necessary information used here for testing schedulers are available online [40].

Huxley Muscle Model was implemented more than 50 years ago, and it is still the most commonly used model for muscle contraction. A DataFlow implementation was developed by Maxeler, achieving the acceleration factor of 33, for array sizes between 32 MB and 384 MB. Total execution time using the CPU varied between 5s and 60s, depending on array size.

N–Body simulation simulates interactions between N particles under gravitational forces in space. Each particle is defined by its mass and its coordinates. Ivan Milankovic developed an Maxeler DataFlow implementation, achieving the speedup of 27.6 times relative to the CPU performance. The algorithm takes 388s on the CPU [45].

The Sequential Monte Carlo method is used for computing posterior distributions. A DataFlow implementation by Thomas Chau et al. [45] achieves a speedup factor of 8.95 comparing relative to the CPU implementation.

A DataFlow implementation for Dense Matrix Multiplication was designed by Maxeler and achieved the speedup of more than 15 for matrix sizes higher than 1000.

Locality Sensitive Hashing is the common technique in data clustering. It solves the nearest neighbor problem and uses high dimension data indexing. A DataFlow implementation designed by Maxeler achieves the speedup of 42 times relative to the CPU implementation.

Table 1 summarizes the acceleration factors of previously mentioned applications using the Maxeler DataFlow hardware. The conditions and

Table 1 The acceleration factors for algorithms implemented using DataFlow hardware.

Research article	Algorithm name	Acceleration factors	Conditions that must hold
Stojanovic et al. [19]	Gross-Pitaevskii	8.2x	–
Korolija et al. [36]	Lattice-Boltzmann	17x	Matrix dimensions 320×112
Kos et al. [37]	Odd-even merge network sort	100–150x	More than 1000 arrays
Stanojevic et al. [44]	Spherical code design	18–24x	$3 <= D <= 6$
Bezanic et al. [39]	RSA	25–30%	Size of file $> 40\,MB$
Maxeler [45]	Huxley muscle model	33x	Array size $> 8\,MB$
Maxeler [45]	N–body simulation	27.6x	–
Thomas Chau et al. [45]	Sequential Monte Carlo	8.95x	–
Maxeler [45]	Dense matrix multiplication	$>15x$	matrix sizes > 1000
Maxeler [45]	Locality sensitive hashing	42x	size $> 10^6$

restrains use to obtain valid results are not contrived, but rather typical conditions, except for the RSA algorithm that can be accelerated only when encrypting relatively big files.

4.2 A scheduling example

In order to test the proposed schedulers, we have chosen applications that can be accelerated using a DataFlow hardware. Table 1 summarizes the algorithms implemented using the Maxeler framework found in the open literature and their DataFlow hardware acceleration factors. For each of these algorithms, we have formed two problems (sizes of input sets) that may differ in the amount of computation. For example, for the Lattice–Boltzmann algorithm, the execution time for two different problems differed by 10 times. This allows us to have an example set of jobs that can be scheduled using a DataFlow hardware, given in Table 2, where time is always measured in seconds. Amounts of main memory (*memCPU*, *memRest*) and DataFlow card memory (*memDFE*) are left out of further consideration without losing generality, since checking each of these constraints is uniform with checking constraints that are included. Neither the main memory, nor the DataFlow card memory poses a constraint in the case of these algorithms. The time needed for initializing jobs for the DataFlow hardware (*tInitDFE*) is usually negligible compared to the time needed for processing, so we will also not consider it.

Table 2 A sample set of jobs for the scheduler.

#	Algorithm	%DFEs (%)	%PCIe (%)	tCPU (s)	tDFE (s)
1	Gross–Pitaevskii	50	0	820	100
2	Odd–Even Merge Network Sort	50	100	1000	10
3	Lattice-Boltzmann	20	0	170	10
4	Spherical Code Design	40	0	1800	100
5	RSA	50	100	250	200
6	Huxley Muscle Model	50	100	60	19
7	N–Body Simulation	50	0	388	14
8	Sequential Monte Carlo	20	0	33	3.7
9	Dense Matrix Multiplication	40	0	240	16
10	Locality Sensitive Hashing	50	100	840	20

Table 3 The best schedule for maximizing time gain.

#	Algorithm	%DFEs (%)	%PCIe (%)	tCPU (s)	tDFE (s)	DFE image
4	Spherical Code Design	40	0	1800	100	1
8	Lattice–Boltzmann	20	0	1700	100	1
1	Gross–Pitaevskii	50	0	820	100	2
2	Odd-Even Merge Network Sort	50	100	1000	10	2
6	Gross–Pitaevskii	50	0	820	100	3
7	Odd-Even Merge Network Sort	50	100	1000	10	3

Table 4 The schedule produced by the proposed scheduler for maximizing time gain.

#	Algorithm	%DFEs (%)	%PCIe (%)	tCPU (s)	tDFE (s)	DFE image
4	Spherical Code Design	40	0	1800	100	1
8	Lattice–Boltzmann	20	0	1700	100	1
9	Spherical Code Design	40	0	180	10	1
1	Gross–Pitaevskii	50	0	820	100	2
2	Odd-Even Merge Network Sort	50	100	1000	10	2
7	Odd-Even Merge Network Sort	50	100	1000	10	3

For demonstrating an example of scheduling, we selected the optimal scheduler to schedule six jobs on the DataFlow hardware. Jobs are formed based on the sample set of jobs from the most left column of Table 3. Table 3 also depicts the best schedule in terms of total execution time of all applications.

Table 4 depicts the schedule produced by the proposed scheduler for maximizing time gain. The first–come–first–serve (FCFS) scheduling technique, starting from the list of jobs that are available for scheduling on the DataFlow hardware, schedules first jobs for the first DFE image, packing as many jobs as there can fit. The schedule is the same as the first six jobs from Table 4, where two jobs are packed per one DFE image. To the best of our knowledge, Maxeler does not have a scheduler for scheduling more than one algorithm per one DFE image. A discussion will follow in the next section.

Table 5 The schedule produced by the proposed scheduler for maximizing throughput.

#	Algorithm	%DFEs (%)	%PCIe (%)	tCPU (s)	tDFE (s)	DFE image
2	Odd-Even Merge Network Sort	50	100	1000	10	1
9	Spherical Code Design	40	0	180	10	1
3	Lattice-Boltzmann	20	0	170	10	2
7	Odd-Even Merge Network Sort	50	100	1000	10	2
4	Spherical Code Design	40	0	1800	100	3
8	Lattice-Boltzmann	20	0	1700	100	3

Table 6 Comparison of scheduling algorithms for a given example.

Scheduling algorithm	TCPUtotal (s)	TDFEtotal (s)	Difference (s)	Speed-up
FCFS	4860	520	4340	9.3461538462
Best schedule for time gain	7140	420	6720	17
Proposed schedule for time gain	6500	330	6170	19.696969697
Proposed schedule for throughput	5850	240	5610	24.375

Table 5 depicts the schedule produced by the proposed scheduler for maximizing throughput. The best schedule that maximizes throughput is the same.

Table 6 presents the following parameters that can be used for comparison between these schedulers: total execution time if all jobs are executed on the CPU ($tCPUtotal$), total execution time if all jobs are executed on the DataFlow hardware ($tDFEtotal$), difference between these two values, and the speed-up factor measured as $tCPUtotal$ divided by $tDFEtotal$.

To keep the pseudo-code of heuristic Algorithm 2 short, we have omitted the limitation in the number of jobs that are taken into consideration for scheduling. If more than 20 jobs are available for scheduling, only the first 20 sorted by the initial score are taken into account. The duration of scheduling only five jobs is around 0.06s executed on Intel i5 650 processor. Scheduling six jobs lasts for around 0.45s.

4.3 Single node scheduling using a synthetic benchmark

Let us define the total acceleration factor as the ratio between the time needed only for the CPU to execute all jobs and the time needed for both the CPU and the DataFlow hardware to execute the same set of jobs. If we vary the amount of CPU jobs and streaming jobs, the total acceleration factor will vary as well. The higher the number of CPU jobs, the smaller the acceleration factor will be. The graph from Fig. 3 depicts acceleration factors depending on the ratio of CPU jobs and streaming jobs. The horizontal axis is scaled linearly and represents the ratio. The value of one corresponds to the ratio between CPU jobs and streaming jobs for which the DataFlow hardware and the CPU are on average equally occupied. Three scheduling policies are included in the simulation. Each one of them uses the same set of jobs. The first scheduling policy optimizes time gain. The second one optimizes throughput. The third one (FCFS) schedules jobs to the DFE in the same order in which the jobs have occurred, packing on each DFE image as many consecutive jobs as possible.

Each benchmark is composed of algorithms presented in Table 1. Duration of jobs varies with sizes of datasets that applications have to process. For each simulation scenario, 100 iterations are performed. We tested each of the scheduling algorithms using 20 ratios of CPU jobs

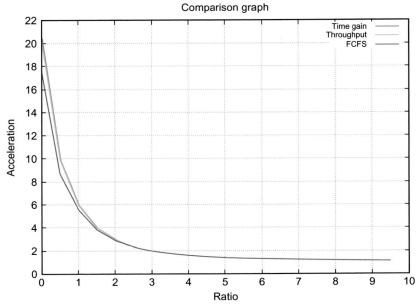

Fig. 3 Acceleration versus the ratio of CPU jobs and streaming jobs.

and streaming jobs (from 0 to 9.5 with the step of 0.5). The exact values of acceleration factors for the graph shown in Fig. 3 are given in Table A1 in the Appendix.

Fig. 4 depicts electrical power reduction factors for all three scheduling policies depending on ratios between jobs. The reduction factor is calculated based on the comparison of CPU and DataFlow hardware execution times for a given set of jobs and power consumption of a desktop workstation with and without the Maxeler MAX2 DataFlow hardware card. The exact values are given in Table A2 in the Appendix.

From the result of comparison between scheduling techniques from Table 6, we can see that the best schedule for the time gain produces the biggest reduction of total execution time. However, the schedule produced by the proposed scheduler for maximizing time gain leaves enough resources on the third DFE image, so that jobs that can come while the first two DFE images are executed can be packed next to the job number 7 from Table 4. Note that this does not mean that the proposed schedule is better than the best schedule, but rather that the best schedule includes only the jobs that are available at the moment of scheduling. The limitation of the best scheduling technique is that it is NP-hard. On the other hand, the FCFS scheduler is rather a queue than a real scheduler.

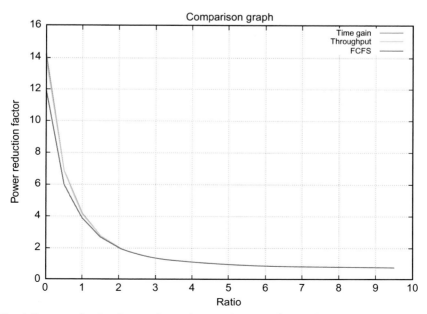

Fig. 4 Power reduction factors depending on the ratio of CPU jobs and streaming jobs.

The example from Table 4 clearly demonstrates the potential flaw of the proposed scheduler for maximizing time gain. The last two jobs in the schedule have acceleration factors of 100. Yet, they are not scheduled within the first DFE image. There are two points of view in this case. The first one is that the last two jobs should be scheduled at the very beginning, even separately on the first two DFE images. This would obviously finish the jobs that are best suited for the DataFlow hardware as soon as possible, achieving a relatively big difference in execution times of these jobs on the CPU and DataFlow hardware. The second point is that these jobs can be left for later, so that they can fit with other jobs that do not require a PCIe bus during execution. The proposed scheduler for maximizing the throughput solves this issue, as shown in Table 5.

Acceleration factors from Table 6 indicate that the performance gain of streaming jobs to DataFlow hardware so that not more than 50% of DataFlow hardware is idle is sufficient to pay off the extra chip die size for the DataFlow hardware.

Fig. 3 depicts the total acceleration factor of the system depending on the scheduling policy. If there are no CPU jobs, the acceleration factor is approximately the same as the ratio between the time needed for the CPU to execute all streaming jobs and the time needed for the DataFlow hardware to execute them. As the amount of CPU jobs increases, the total acceleration factor decreases. Once the amount of CPU jobs increases so that all streaming jobs can be scheduled using DataFlow hardware, there is no more difference between the first two scheduling algorithms. By further increasing the amount of CPU jobs, the acceleration factor decreases, approaching nearly the value of one, which indicates that there is no acceleration.

Fig. 4 depicts the total power consumption reduction factor for each scheduling algorithm depending on the ratio of CPU jobs and streaming jobs. We can see from the precise values given in Table A2 in the Appendix that power consumption is reduced when streaming jobs represent at least one sixth of the total number of jobs.

The scheduler that optimizes time gain proved to be more efficient than the one that optimizes throughputs for all ratios of CPU jobs to streaming jobs. By calling the first two schedulers using the zero ratio of CPU jobs to streaming jobs for 40 times, we have calculated the correlation of 0.952 and the covariance of 12.4. This correlation is expected, since both schedulers accelerate the same set of jobs.

The important finding is that after the ratio becomes equal or higher than one, the difference between total execution time using the first two schedulers becomes <1% of execution time. This emphasizes the importance of the ratio between the two types of jobs. In the case of FCFS scheduling, the difference in total execution times becomes <1% only when the ratio reaches or exceeds three.

By varying the percentages of main memory needed for streaming jobs, we can simulate the conditions where main memory becomes the bottleneck. As a result, additional constraints appear for the scheduler when searching for the next job to schedule on a DFE image. As expected, results indicate lower acceleration factors, as well as smaller difference between total execution times using the two proposed schedulers.

Compared to the scheduler based on mixed integer linear programming proposed by Abdessamad et al. [27], our schedulers are also based on the branch and bound algorithm, but they consider not only the cost of scheduling jobs onto the FPGA, but also the constraints, allowing a job that is not likely to produce the best result to be placed on the same DFE image besides other jobs, if there is no better job that can be placed instead. This makes our scheduler slower, but it allows for reduction in total execution time, as we can see in Table 4, where job number 9 is packed onto the first DFE image, even though putting it onto DataFlow hardware reduces execution time the least. This reduces the total computation time far more than the scheduling time. However, our schedulers do not schedule even close to 50 jobs per scheduling, but rather only the estimated best jobs that can fit on up to few DFE images, leaving scheduling other available jobs for later, closer to their execution, as new jobs might arrive in the meantime. Note that Abdessamad et al. [25] scheduled tasks (i.e., jobs) according to the task dependency graph, leaving only part of tasks available for scheduling at any given moment. The main reason we can schedule fewer jobs is that DataFlow hardware is usually capable of running only up to few algorithms efficiently. Increasing the number of jobs that can be executed simultaneously would bring more constraints for schedulers, leading to slower scheduling.

The scheduler proposed by Fei et al. [32] is in some parts similar to our schedulers. However, our solution is designed to be fast and adaptable in the runtime.

In the case of Maxeler's default scheduler, to the best of our knowledge, just a single job is run on a single DFE at any time.

The multi-agent optimization algorithm proposed by Zheng and Wang [46] leads to the optimal solution by iterating towards the best candidate solution. We found this algorithm to be computationally demanding in the case of having enough agents to cover all the possibilities that arise with adding various constraints (CPU memory, DataFlow hardware memory, CPU computation time, DataFlow hardware computation time, FPGA limits, PCIe constraints, etc.). Instead, our heuristic approach fits jobs onto the first DFE image relatively fast and efficiently compared to the best schedule and FCFS schedule scenarios.

Hamilton et al. [28] developed a system for mixed-architecture processes supporting context switching and blocking. However, their execution model does not consider DataFlow with all the constraints the DataFlow architecture brings.

The proposed schedulers can be configured to support more than one DataFlow hardware and more than one CPU. This way, configuration prefetching can be used as an effective mechanism for hiding the delay of preparing DFE images. In prefetching, a DataFlow hardware job can be loaded as soon as there is free DataFlow hardware, and it is known that jobs will be processed, hiding the configuration delay. Khuat et al. [29] reported the overall reduction in execution time of 22% using their prefetching mechanism. Note that in order for prefetching to be possible, the DataFlow hardware must not be occupied before the arrival of jobs.

4.4 Cluster scheduling using a synthetic benchmark

The cluster scheduling algorithm is tested against scheduling jobs on cluster nodes only based on (1) the availability of resources determined by a single variable calculated as a function of all available resources on the node and (2) the average amount of jobs on nodes. In both cases, for all tested ratios between DataFlow and control-flow jobs, from all jobs available for scheduling, jobs are initially only scheduled until the average resource availability on nodes approaches approximated expected node occupation. A node occupation is calculated based on node capability, the number of nodes, and the number of jobs given to the scheduler per unit of time. This way, if there are too many jobs to fit onto nodes at the scheduling time, only a portion of jobs will actually be scheduled, avoiding waiting of jobs at certain nodes while other nodes are not occupied at all.

At the node level, our heuristic cluster scheduler is tested against the best schedules, where only the best 10 jobs would be considered by the scheduler due to NP complexity.

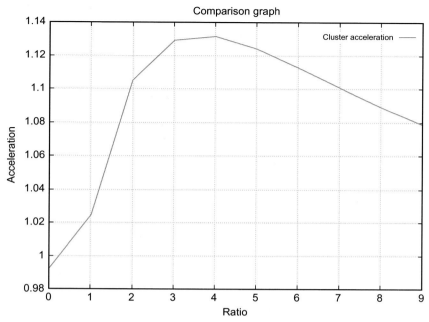

Fig. 5 Acceleration factors of the proposed cluster scheduler.

Fig. 5 depicts acceleration factors compared to the scheduler that sends jobs to cluster nodes without taking into the account the relation between specific job resource requirements and availability of each type of resource on each node, except that a node must have enough resources for a job.

As it can be seen from the figure, acceleration is even somewhat worse for the ratio of jobs such that there are no jobs that have to be scheduled on the control-flow processor. This can be explained by using the best schedule in the case of scheduling for the cluster where the availability of resources is determined by only one variable. In case there are approximately the same amounts of jobs for control-flow and DataFlow hardware, the acceleration obtained by the proposed cluster scheduler can be neglected. However, once there are more jobs for control-flow hardware, the proposed cluster scheduler distributes jobs on cluster nodes more equally in terms of respecting each resource separately, leading to acceleration factors up to 1.13. As the DataFlow jobs are becoming rare compared to control-flow jobs, the acceleration factor tends to fall to the value of one. If there are no DataFlow jobs at all, both schedulers work exactly the same, while the DataFlow processor is able to finish all jobs for the time that the control-flow processor needs for finishing the jobs that can be executed only on the control-flow processor.

4.5 On the validity of results

The most important threat to validity is the fact that it is not possible to know what profile of applications will be mostly executed on a cluster architecture. This work is based on the assumption that computationally demanding algorithms, sometimes with several dimensions of input data, will be predominantly executed, since DataFlow hardware enables the execution of more instructions per second, and computation in contemporary HPC applications tends to be performed throughout a physical material or environment [38,47].

It is hard to know execution time of an application before runtime. Fortunately, DataFlow hardware is often programmed for processing a certain amount of data which is known at the time of starting the job, and therefore, the duration of the job can be calculated prior to the execution. However, the results of this work are based on estimated durations of execution of algorithms, and it is not possible to know the durations of execution of algorithms for input data sizes that will be used in the future. This means that the ratio of control-flow and DataFlow hardware in clusters has to follow needs for these types of processing in the future.

Although DataFlow hardware might have more memory than the CPU, our benchmark set consisted of applications which do not require a lot of memory. A possible threat to validity is the fact that streaming jobs might require a relatively high amount of main memory during the execution on DataFlow hardware, preventing scheduling them at any moment, which would reduce the total acceleration factor.

5. Conclusions

While control-flow processors offer high flexibility, their high throughput computing can be achieved only by increasing the number of cores. On the other side, the DataFlow paradigm offers higher instruction per watt throughput, as well as better performances when executing specific high performance computing applications [1]. As a consequence, some computer clusters and clouds consist of both types of processors. This imposes the problem of scheduling job execution on hybrid control-flow/ DataFlow architectures. This article discussed related scheduling algorithms from various fields, including scheduling in heterogeneous computing architectures, scheduling real-time jobs, and resource-constrained project scheduling. Their advantages and drawbacks are also discussed.

We further propose a branch-and-bound based scheduler for runtime scheduling of multiple jobs per DataFlow element. Unlike most schedulers available in the open literature, this scheduler considers scheduling only a relatively small number of jobs in a greedy manner, where each job can have a noticeable impact on total execution time depending on whether it is executed using a CPU or DataFlow hardware. For this reason, the proposed scheduler is more than an order of magnitude faster than those based on genetic algorithms and mixed integer linear programming, while being capable of making schedules which are relatively close to the best possible schedules.

Compared to the existing Maxeler's default scheduler, which dedicates a complete DataFlow hardware element per job, the advantage of the proposed scheduler is that it enables sharing DataFlow hardware element resources among multiple jobs.

The cluster scheduler schedules jobs onto cluster nodes based on job requirements for specific resources, as well as the average ratio between available resources on nodes. The proposed node scheduler schedules the execution of jobs on a single DataFlow hardware element of each node.

Future work will include priority scheduling algorithms, so that important jobs can utilize DataFlow hardware resources prior to scheduling low priority jobs. Possible future work includes developing a real-time scheduling algorithm, as DataFlow jobs duration can often be calculated in advance.

Algorithm 1. Moving average kernel.

```
package chap01_overview.ex2_movingaverage;
import com.maxeler.maxcompiler.v1.kernelcompiler.Kernel;
import com.maxeler.maxcompiler.v1.kernelcompiler.KernelParameters;
import com.maxeler.maxcompiler.v1.kernelcompiler.types.base.HWVar;
public class MovingAverageOverviewKernel extends Kernel {
    public MovingAverageOverviewKernel(KernelParameters parameters,
    int N) {
        super(parameters);
        HWVar x = io.input("x", hwFloat(8, 24)); // Input
        HWVar x_prev = stream.offset(x, -1); // Data
    HWVar x_next = stream.offset(x, 1);
        HWVar cnt = control.count.simpleCounter(32, N); // Control
        HWVar sel_nl = cnt > 0;
        HWVar sel_nu = cnt < N-1;
```

```
        HWVar sel_m = sel_nl & sel_nu;
        HWVar prev = sel_nl ? x_prev : 0;
        HWVar next = sel_nu ? x_next : 0;
        HWVar divisor = sel_m ? constant.var(hwFloat(8, 24), 3) : 2;
        HWVar sum = prev + x + next;
        HWVar result = sum/divisor;
        io.output("y", result, hwFloat(8, 24));
    }
}
```

Algorithm 2. Heuristic algorithm for finding the best schedule.

```
J0 = {J0_1, J0_2,...J0_n} // available jobs for scheduling
I = {} // candidate set of DFE images whose elements are set of jobs
nI = 0 // number of DFE images used by jobs in the candidate schedule
B = {} // the best schedule found so far
find_best(k) { // recursion with maximum depth of k
  if k=0
    then{
    if score(I) > score(B) // if current score is better than the best found so far
      then B = I
    }else{
    for all jobs j in J0 {
      J0 = J0 - j // exclude job j from jobs
                  // available for scheduling
      if( (PCIe(j) + PCIe(I[ni]) < 1) && // if j can fit on
                                 // current DFE image
         (DFE(j) + DFE(I[nI]) < 1) ) && // 1 represents 100%
         (mem(j, I[nI])) ) {
      I[nI] = I[nI] + j // add it to the current DFE image
      find_best(k-1) // continue searching until given depth
      I[nI] = I[nI] - j // remove j from current DFE image
                        // and try other jobs
      }
      // try putting j on a new DFE image:
      nI = nI + 1
      I[nI] = I[nI] + j
      find_best(k-1) // continue searching until given depth
```

```
    I[nI] = I[nI] − j // remove j from current DFE image
    nI = nI − 1 // job j was the last job on the DFE image
    J0 = J0 + j // return job j to the set of jobs
                // available for scheduling
  }
 }
}
```

Algorithm 3. Scheduling execution of jobs.

```
if (currentScore > bestScore) {
  bestScore = currentScore;
  bestSchedule = currentSchedule;
  bestScheduleFPGAs.clear();
  for (auto &it : currentSchedule)
    bestScheduleFPGAs.push_back(it->_FPGAimage);
}
if(currentSchedule.size() == MAX_SCHED_JOBS) return;
// Calculate occupation of the nFPGAimage
for (auto &it : currentSchedule)
  if( it->_FPGAimage==nFPGAimage )
    it->sum(sumPercentageDFEs, sumPercentagePCIe, sumTInitDFEs,
        sumMemDFEs, sumTDFE, sumTCPU, tDFEmax);
// Adding each job separately to the currentSchedule
for (auto &it : jobs) { // for each job
  if(it->add(&currentSchedule, sumPercentageDFEs,
        sumPercentagePCIe,
        sumMemDFEs, sumTDFE, sumTCPU,
        sumTInitDFEs, tDFEmax, nFPGAimage) ){
  score1 += it->score(); // increasing by the it job score
  double scoreForMode;
  if(mode == THROUGHPUT)
    scoreForMode = 1.0*sumTCPU/tDFEmax;
  else // (mode == TIMEGAIN) || (mode == FCFS)
    scoreForMode = sumTCPU-tDFEmax;
  // If no other job can fit on nFPGAimage
  if(cantFit(jobs, sumPercentageDFEs, sumPercentagePCIe,
        sumMemDFEs, tDFEmax))
```

```
    // Schedule the rest of jobs on nFPGAimage + 1
    schedJobs(mode, jobs, currentSchedule,
            currentScore + factor*(score1 + scoreForMode),
            bestScore, nFPGAimage + 1, factor/1.1);
  else
    // Schedule on the same DFE image
    schedJobs(mode, jobs, currentSchedule,
            currentScore + factor*score1,
            bestScore, nFPGAimage, factor);
  if(mode!=FCFS)
    it->remove(&currentSchedule, sumPercentageDFEs,
            sumPercentagePCIe, sumMemDFEs,
            sumTDFE, sumTCPU, sumTInitDFEs,
            tDFEmax, nFPGAimage);
  }else{
  if(mode==FCFS)
    // Schedule the rest of jobs on nFPGAimage + 1
    schedJobs(mode, jobs, currentSchedule, currentScore + 1,
            bestScore, nFPGAimage + 1, factor);
  }
}
```

Acknowledgments

This work has been partially funded by the Ministry of Education and Science of the Republic of Serbia (TR32047 and III44006). We would also like to thank prof. Roberto Giorgi for the possibility to develop scheduling algorithm for Decoupled Threaded Architecture (DTA), supported by the HiPEAC project.

 Appendix

Table A1 Total execution time depending on ratio of streaming jobs and jobs for the CPU.

Ratio between	Time gain	Throughput	FIFO
0	20.794485	20.022567	17.421574
0.5	10.057943	9.814917	8.689677
1	6.045375	5.982859	5.604964
1.5	4.031196	4.008519	3.866492

Table A1 Total execution time depending on ratio of streaming jobs and jobs for the CPU.—cont'd

Ratio between	Time gain	Throughput	FIFO
2	2.978221	2.968537	2.906722
2.5	2.379296	2.374582	2.344153
3	2.011726	2.009187	1.992688
3.5	1.771775	1.770297	1.760648
4	1.607195	1.606279	1.600286
4.5	1.489711	1.489115	1.485207
5	1.403059	1.402655	1.400003
5.5	1.337390	1.337107	1.335246
6	1.286472	1.286268	1.284925
6.5	1.246218	1.246067	1.245074
7	1.213856	1.213742	1.212992
7.5	1.187457	1.187370	1.186793
8	1.165646	1.165578	1.165128
8.5	1.147421	1.147367	1.147011
9	1.132040	1.131996	1.131710
9.5	1.118940	1.118905	1.118673

Table A2 Power reduction factors depending on ratio of streaming jobs and jobs for the CPU.

Ratio between	Time gain	Throughput	FIFO
0	14.396181	13.861777	12.061089
0.5	6.963191	6.794942	6.015930
1	4.185259	4.141979	3.880359
1.5	2.790828	2.775128	2.676802
2	2.061845	2.055141	2.012346
2.5	1.647204	1.643941	1.622875
3	1.392733	1.390975	1.379553

Continued

Table A2 Power reduction factors depending on ratio of streaming jobs and jobs for the CPU.—cont'd

Ratio between	Time gain	Throughput	FIFO
3.5	1.226613	1.225590	1.218910
4	1.112673	1.112039	1.107890
4.5	1.031338	1.030925	1.028220
5	0.971348	0.971068	0.969232
5.5	0.925885	0.925689	0.924401
6	0.890634	0.890493	0.889563
6.5	0.862766	0.862661	0.861974
7	0.840361	0.840282	0.839763
7.5	0.822085	0.822025	0.821625
8	0.806985	0.806938	0.806627
8.5	0.794368	0.794331	0.794084
9	0.783720	0.783689	0.783491
9.5	0.774650	0.774626	0.774465

References

[1] V. Milutinović, J. Salom, N. Trifunovic, R. Giorgi, The Dataflow Paradigm, In Guide to Dataflow Supercomputing, Springer, Cham, 2015, pp. 1–39.
[2] A.R. Hurson, M.K. Krishna, Dataflow Computers: Their History and Future, Wiley Encyclopedia of Computer Science and Engineering, 2007.
[3] C. Arvind, Dataflow architectures, Ann. Rev. Comput. Sci. (1986) 225–253.
[4] A.H. Veen, Dataflow machine architecture, ACM Comput. Surv. 18 (4) (1986) 365–396.
[5] B. Lee, A.R. Hurson, Issues in dataflow computing, Adv. Comput. 37 (1993) 285–333.
[6] B. Lee, A.R. Hurson, Dataflow architectures and multithreading, Computer 8 (1994) 27–39.
[7] A. Milenkovic, V. Milutinovic, Cache injection: A novel technique for tolerating memory latency in bus-based SMPs, in: European Conference on Parallel Processing, Springer, Berlin, Heidelberg, 2000, pp. 558–566.
[8] A. Grujic, M. Tomasevíc, V. Milutinovic, A simulation study of hardware-oriented DSM approaches, IEEE Parallel Distrib. Technol. Syst. Appl. 4 (1) (1996) 74–83.
[9] D. Milutinovic, V. Milutinovic, B. Soucek, The honeycomb architecture, Computer 20 (1987) 81–83.
[10] H. Kwak, B. Lee, A.R. Hurson, S.H. Yoon, W.J. Hahn, Effects of multithreading on cache performance, IEEE Trans. Comput. 48 (2) (1999) 176–184.
[11] V. Milutinovic, Surviving the design of a 200 MHz risc microprocessor: lessons learned, IEEE Comput. Soc. (1996).

[12] P. Knezevic, B. Radunovic, N. Nikolic, T. Jovanovic, D. Milanov, M. Nikolic, V. Milutinovic, S. Casselman, J. Schewel, The architecture of the Obelix-an improved Internet search engine, in: System Sciences, Proceedings of the 33rd Annual Hawaii International Conference on, IEEE, January, 2000, p. 11.

[13] W.M. Johnston, J.R.P. Hanna, R.J. Millar, Advances in dataflow programming languages, ACM Comput. Surv. 36 (1) (2004) 1–34.

[14] O. Pell, O. Mencer, K.H. Tsoi, W. Luk, Maximum performance computing with dataflow engines, in: W. Vanderbauwhede, K. Benkrid (Eds.), High-Performance Computing Using FPGAs, Springer-Verlag, 2013.

[15] M.J. Flynn, O. Mencer, V. Milutinovic, G. Rakocevic, P. Stenstrom, R. Trobec, M. Valero, Moving from petaflops to petadata, Commun. ACM 56 (5) (2013) 39–42.

[16] V. Milutinovic, M. Kotlar, M. Stojanovic, I. Dundic, N. Trifunovic, Z. Babovic, DataFlow Systems: From Their Origins to Future Applications in Data Analytics, Deep Learning, and the Internet of Things, in: DataFlow Supercomputing Essentials, Springer, Cham, 2017, pp. 127–148.

[17] N. Trifunovic, B. Perovic, P. Trifunovic, Z. Babovic, A.R. Hurson, A novel infrastructure for synergistic dataflow research, development, education, and deployment: the Maxeler AppGallery project, in: Advances in Computers, vol. 106, Elsevier, 2017, pp. 167–213.

[18] M. Harchol-Balter, Performance Modeling and Design of Computer Systems: Queueing Theory in Action, Cambridge University Press, 2013.

[19] S. Stojanović, D. Bojić, V. Milutinović, Solving gross Pitaevskii equation using dataflow paradigm, The IPSI BgD Trans. Int. Res. 9 (2) (2013) 17–22.

[20] V. Blagojević, D. Dragan, M. Bojović, M. Cvetanović, J. Đorđević, Đ. Đurđević, B. Furlan, et al., A systematic approach to generation of new ideas for PhD research in computing, in: Advances in computers, vol. 104, Elsevier, 2017, pp. 1–31.

[21] I. Ahmad, Y.K. Kwok, On parallelizing the multiprocessor scheduling problem, IEEE Trans. Parallel Distrib. Syst. 10 (4) (1999) 414–431.

[22] D.G. Amalarethinam, G.J. Mary, Modified dynamic scheduling algorithm for multiprocessor system, Int. J. Eng. Technol. 2 (4) (2009).

[23] K. Ramamritham, A.J. Stankovic, Efficient scheduling algorithms for real-time multiprocessor systems, IEEE Trans. Parallel Distrib. Syst. 1 (2) (1990) 184–194.

[24] B. Lee, A.R. Hurson, T.Y. Feng, A vertically layered allocation scheme for data flow systems, J. Parallel Distrib. Comput. 11 (3) (1991) 175–187.

[25] B. Hamidzadeh, D.J. Lilja, Self-adjusting scheduling: An on-line optimization technique for locality management and load balancing, in: 1994 International Conference on Parallel Processing, vol. 2, IEEE, 1994, pp. 39–46.

[26] Y. YuHai, Y. Shengsheng, B. XueLian, A new dynamic scheduling algorithm for real-time heterogeneous multiprocessor systems, In Workshop on Intelligent Information Technology Application (IITA), IEEE 2007 (2007) 112–115.

[27] A.E.C. Abdessamad, S. Omar, B.A. Rabie, B. Nicolas, A. Abdelhakim, Mathematical programming models for scheduling in a CPU/FPGA architecture with heterogeneous communication delays, J. Intell. Manuf. 1–12 (2015). Springer US.

[28] B.K. Hamilton, M. Inggs, H.K.-H. So, Scheduling mixed-architecture processes in tightly coupled FPGA-CPU reconfigurable computers, in: IEEE 22nd Annual International Symposium on Field-Programmable Custom Computing Machines (FCCM), 2014, p. 240.

[29] Q.-H. Khuat, D. Chillet, M. Hubner, Considering reconfiguration overhead in scheduling of dependent tasks on 2D reconfigurable FPGA, in: NASA/ESA Conference on Adaptive Hardware and Systems (AHS), 2014, pp. 1–8.

[30] A. Cilardo, E. Fusella, L. Gallo, A. Mazzeo, Joint Communication Scheduling and Interconnect Synthesis for FPGA-Based Many-Core Systems, Design, Automation and Test in Europe Conference and Exhibition (DATE), 2014, pp. 1–4.

[31] X. Iturbe, K. Benkrid, C. Hong, A. Ebrahim, T. Arslan, I. Martinez, Runtime scheduling, allocation, and execution of real-time hardware tasks onto xilinx FPGAs subject to fault occurrence, Int. J. Reconfig. Comput. 2013 (2013) 1–32. Article ID 905057.

[32] T. Fei, Z. Lin, L. Yuanjun, Job Shop Scheduling with FPGA-Based F4SA, Configurable Intelligent Optimization Algorithm, Part of the series Springer Series in Advanced Manufacturing, Springer International Publishing, 2015, pp. 333–347.

[33] S. Van de Vonder, E. Demeulemeester, W. Herroelen, A classification of predictive-reactive project scheduling procedures, J. Schedul. 10 (3) (2007) 195–207.

[34] R.H. Möhring, F. Stork, Linear preselective policies for stochastic project scheduling, Math. Methods Oper. Res. 52 (3) (2000) 501–515.

[35] T. Yu, B. Feng, M. Stillwell, L. Guo, Y. Ma, J. Thomson, Lattice-based scheduling for multi-FPGA systems, in: 2018 International Conference on Field-Programmable Technology (FPT), IEEE, 2018, pp. 318–321.

[36] N. Korolija, J. Popović, M. Cvetanović, M. Bojović, Dataflow-based parallelization of control-flow algorithms, Adv. Comput. 104 (2017) 73–124.

[37] A. Kos, V. Rankovic, S. Tomazic, Sorting networks on Maxeler dataflow super-computing systems, Adv. Comput. 96 (2015) 139–186.

[38] V. Milutinović, N. Trifunović, N. Korolija, J. Popović, D. Bojić, Accelerating program execution using hybrid control flow and dataflow architectures, in: Telecommunication Forum (Telfor), 2017 25th, IEEE, November, 2017, pp. 1–4.

[39] N. Bezanic, J. Popovic-Bozovic, V. Milutinovic, I. Popovic, Implementation of the RSA algorithm on a dataflow architecture, Trans. Int. Res. 9 (2) (2013) 11–16.

[40] R. Roma, M. Sidia, H.P. Tan, Design and analysis of a class-aware recursive loop scheduler for class-based scheduling, Elsevier 63 (9–10) (2006) 839–863.

[41] J.D. Anderson, J. Wendt, Computational Fluid Dynamics, McGraw-Hill, New York, 1995.

[42] S. Chen, G.D. Doolen, Lattice Boltzmann method for fluid flows, Ann. Rev. Fluid Mech. (1998) 329–364.

[43] P.L. Bhatnagar, E.P. Gross, M. Krook, A model for collision processes in gases: I. Small amplitude processes in changed and neutral one-component system, Phys. Rev. 94 (1954) 511–525.

[44] I. Stanojevic, V. Senk, V. Milutinovic, Application of Maxeler dataflow super-computing to spherical code design, Trans. Int. Res. 9 (2) (2013) 1–4.

[45] appgallery. www.appgallery.maxeler.com, web site visited on January 2019. 2019.

[46] X.L. Zheng, L. Wang, A multi-agent optimization algorithm for resource constrained project scheduling problem, Exp. Syst. Appl. 42 (15–16) (2015) 6039–6049.

[47] J. Beal, M. Viroli, Space-time programming, Philos. Trans. R. Soc. Lond. A Math. Phys. Eng. Sci. 373 (2015) 2046.

About the authors

Nenad Korolija is with the faculty of the School of Electrical Engineering, University of Belgrade, Serbia. He received a PhD degree in electrical engineering and computer science in 2017. His interests and experiences include developing software for high performance computer architectures and Dataflow architectures. During 2008, he worked on the HIPEAC FP7 project at the University of Siena, Italy. In 2013, he was an intern at the Google Inc., Mountain View, California, USA. In 2017, he worked for Maxeler Ltd., London. During 2021, he worked for Johns Hopkins University on parallelizing the protein formation simulator.

Dragan Bojić received a Ph.D. degree in Electrical Engineering and Computer Science from the University of Belgrade in 2001. He is a professor at the School of Electrical Engineering, University of Belgrade. His research interests include parallel computing, software engineering techniques and tools, and artificial intelligence.

Ali R. Hurson is a professor of Electrical and Computer Engineering at Missouri S\&T. For the period of 2008–20012 he served as the computer science department chair. Before joining Missouri S\&T, he was a professor of Computer Science and Engineering department at The Pennsylvania State University. His research for the past 40 years has been supported by NSF, DARPA, Department of Education, Air Force, Office of Naval Research, NCR Corp., General Electric, IBM, Lockheed Martin, Penn State University, and Missouri S&T. He has published over 350 technical papers in areas including database systems, multidatabases, global information sharing processing, application of mobile agent technology, object-oriented databases, Mobile and pervasive computing, computer architecture and cache memory, parallel and distributed processing, Dataflow architectures, and VLSI algorithms. Professor Hurson served as an IEEE and ACM distinguish speaker.

Prof. Veljko Milutinović (1951) received his PhD from the University of Belgrade in Serbia, spent about a decade on various faculty positions in the USA (mostly at Purdue University and more recently at the University of Indiana in Bloomington), and was a co-designer of the DARPAs pioneering GaAs RISC microprocessor and the related GaAs Systolic Array with about 14000 GaAs microprocessors. Later, for almost three decades, he taught and conducted research at the University of Belgrade, in EE, MATH, BA, and PHYS/CHEM. His research is mostly in datamining algorithms and DataFlow computing, with the emphasis on mapping of data analytics algorithms onto fast energy efficient architectures. Most of his research was done in cooperation with industry (Intel, Fairchild, Honeywell, Maxeler, HP, IBM, NCR, RCA, etc...). For 10 of his books, forewords were written by 10 different Nobel Laureates with whom he cooperated on his past industry sponsored projects. He published 40 books (mostly in the USA), he has over 100 papers in SCI journals (mostly in IEEE and ACM journals), and he presented invited talks at over 400 destinations worldwide. He has well over 1000 Thomson-Reuters WoS citations, well over 1000 Elsevier SCOPUS citations, and about 4000 Google Scholar citations. He is a Life Fellow of the IEEE and a Member of The Academy of Europe and a Foreign Member of The Montenegro National Academy of Sciences and Arts.